T0225760

Network and Communication

Center for Electronics and Information Studies,
Chinese Academy of Engineering

Network and Communication

Research on the Development of Electronic
Information Engineering Technology
in China

Center for Electronics and Information Studies,
Chinese Academy of Engineering
Beijing, China

ISBN 978-981-15-4595-5   ISBN 978-981-15-4596-2   (eBook)
https://doi.org/10.1007/978-981-15-4596-2

Jointly published with Science Press, Beijing, China

The print edition is not for sale in China (Mainland). Customers from China (Mainland) please order
the print book from: Science Press.

The English book is translated on the basis of topic 6 of *Research on the development of electronic
information engineering technology in China (2018-2019)* published by Science Press.

This Springer imprint is published by the registered company Springer Nature Singapore Pte Ltd.
The registered company address is: 152 Beach Road, #21-01/04 Gateway East, Singapore 189721,
Singapore

Preface

The *Research on the Development of Electronic Information Engineering Technology in China* Book Series

In today's world, the wave of information technologies featured by digitalization, networking, and intelligence is gaining momentum. Information technologies are experiencing rapid changes with each passing day and are fully applied in production and life, bringing about profound changes in global economic, political, and security landscapes. Among diverse information technologies, electronic information engineering technology is one of the most innovative and widely used technologies and plays its greatest role in driving the development of other S&T fields. It is not only a field of intense competition in technological innovation, but also an important strategic direction for key players to fuel economic growth and seek competitive advantages over other players. Electronic information engineering technology is a typical "enabling technology" that enables technological progress in almost all other fields. Its integration with biotechnology, new energy technology, and new material technology is expected to set off a new round of technological revolution and industrial transformation, thereby bringing about new opportunities for the evolution of human society. Electronic information is a typical "engineering technology" and one of the most straightforward and practical tools. It realizes direct and close integration of scientific discoveries and technological innovations with industrial developments, greatly speeding up technological progress. Hence, it is regarded as a powerful force to change the world. Electronic information engineering technology is a vital driving force of China's rapid economic and social development in the past seven decades, especially in the past four decades of reform and opening up. Looking ahead, advances and innovations in electronic information engineering technology will remain to be one of the most important engines driving human progress.

Chinese Academy of Engineering (CAE) is China's foremost academic and advisory institution in engineering and technological sciences. Guided by the general development trends of science and technology around the world, CAE is committed to providing scientific, forward-looking, and timely advice for innovation-driven scientific and technological progress from a strategic and long-term perspective. CAE's mission is to function as a national high-end think tank. To fulfill the mission, the Division of Information and Electronic Engineering, under the guidance of its Vice President Zuoning Chen, Director Xicheng Lu, and the Standing Committee, mobilized more than 300 academicians and experts to jointly compile the General Section and the Special Themes of this book (hereinafter referred to as the *Blue Book*). The first stage of compilation was headed by Academicians Jiangxing Wu and Manqing Wu (from the end of 2015 to June 2018), and the second one was headed by Academicians Shaohua Yu and Jun Lu (since September 2018). The purposes of compiling the *Blue Book* are:

By analyzing technological progress and introducing major breakthroughs and marked achievements made in the electronic information field both at home and abroad each year to provide reference for China's scientific and technical personnel to accurately grasp the development trend of the field and provide support for China's policymakers to formulate related development strategies.

The *Blue Book* is compiled according to the following principles:

1. Ensure appropriate description of annual increment. The field of electronic information engineering technology enjoys a broad coverage and high development speed. Thus, the General Section should ensure an appropriate description of the annual increment, which is about the recent progress, new characteristics, and new trends.
2. Selection of hot points and highlight points. China's technological development is still at a mixed stage where it needs to assume the role of follower, contender, and leader simultaneously. Hence, the Special Themes should seek to depict the developmental characteristics the industry it focuses on and should center on the "hot points" and "highlight points" along the development journey.
3. Integration of General Section and Special Themes. The program consists of two sections: the General section and the Special Themes. The former adopts a macro perspective to discuss the global and Chinese development of electronic information engineering technology, and its outlook; the latter provides detailed descriptions of hot points and highlight points in the 13 subfields.

Application System

8. Underwater acoustic engineering

13. Computer application

Acquiring Perception	**Computation and Control**	**Cyber Security**
3. Sensing 5. Electromagnetic space	10. Control 11. Cognition 12. Computer systems and software	6. Network and communication 7. Cybersecurity

Common Basis

1. Microelectronics and Optoelectronics 2. Optical engineering
4. Measurement, metrology and Instruments
9. Electromagnetic field and electromagnetic environment effect

Classification Diagrams of 13 Subfields of information and electronic engineering technology

The above graphic displays 5 categories and 13 subcategories or special themes that bear distinct granularity. However, every subfield is closely connected with each other in terms of technological correlations, which allows easier matching with their corresponding disciplines.

Currently, the compilation of the *Blue Book* is still at a trial stage where careless omissions are unavoidable. Hence, we welcome comments and corrections.

The Special Theme *Network and Communication* in *Research on the Development of Electronic Information Engineering Technology in China* Book Series

Currently, the entire world is participating in the process of digitalizing, networking, and intellectualizing the deep integration of the network information world with the natural world and human society. This process will lead to a significant improvement of the quality of productivity and introduce new methods of production and lifestyles. This advancement will have a profound and significant impact, similar to the agricultural and industrial civilizations experienced in human history. This process is profoundly changing competition throughout the world. Security, economics, society, military, and culture trends generate new opportunities for national development, new spaces for people's lives, new fields of social governance, and new momentum for industrial upgrading and international competition. Over the next 20 years, the development direction of network communication technology will be human-network-thing three-domain interconnection and its systematic integration with various industries and regions. This direction will be realized by digitalization, networking, and intellectualization leading to the connection, extension, and penetration of the natural world and human society, and the network space will be continuously enriched and expanded. With the acceleration of the deployment of the Fifth Generation Mobile Communication System(5G), the number of network connections has increased from 10 to 100 billion and one trillion, from the interconnection of human-network two-domain to human-network-thing three-domain and multiple interconnections, and from plane interconnections on the ground to three-dimensional interconnections in space and interstellar interconnections (the information network, which is equivalent to adding a nervous layer to the natural world and human society; its essence is deep informationization). The network has the ability to perceive, transmit, store, and process beyond imagination. Combined with traditional industries and regions, this network will create infinite possibilities. Accelerating the construction of "cyber power" in China is a strategic choice to seize the great opportunities of the information revolution and satisfies the urgent need to build a strong country in regard to science and technology, build cyber power, and reshape its international competitiveness.

This Special Theme of *Blue Book* focuses on the network and communication technologies. Global trends, network, and communication development in China. Based on the above consideration, the book is mainly divided into three parts, six chapters.

In the first part, this book summarizes the global development situation and the characteristics of the network and communication, the current situation of our country, its global status, and its future prospects from the three dimensions of academia, technology, and industry. Secondly, from the two aspects of technology and application, three basic technology categories in the field of network communication are selected: mobile communication, data communication, and optical fiber communication. The mobile internet, Internet of Things, edge computing, and quantum communication are the seven major application services. The latest technological progress, future development trend, technology and industry development hot spots and highlights are described. Lastly, 15 hot words in the field of network communication in technology and industrial development are summarized, basic definitions are provided, and the development level of related applications is introduced.

Experts from the China Information and Communication Technologies Group Co., Ltd. (Wuhan Research Institute of Posts and Telecommunications); China Academy of Information and Communications Technology; China Telecom Group Co., Ltd.; China Mobile Communications Group Co., Ltd.; China United Network Communications Group Co., Ltd.; Tsinghua University; Fudan University; Peking University; ZTE; Huawei Technology; Shanghai Jiao Tong University; China Electronics Technology Group Corporation; Beijing University of Posts and Telecommunications; Telecommunications Science and Technology Research Institute, Southeast University; PLA Strategic Support Force Information Engineering University; and University of Electronic Science and Technology of China participated in this research. I express my deep respect and heartfelt thanks.

Due to the large content of network communication and the limited space, this subfield primarily involves mobile communication, data communication, optical fiber communication, mobile internet, and Internet of Things-related technologies. Cyber security is an important aspect in the field of networks and has been introduced in its subfields. Additional content (such as radio and television networks) will gradually be added and improved in follow-up work.

Beijing, China   Shaohua Yu
February, 2020

Acknowledgements

We extend thanks to the academician Shaohua Yu, who is the head of the Strategic Research Group of the Network and Communication Subject Group, and the academician Hequan Wu and Professor Shumin Cao, who are deputy heads. The specific work was led by the academician Shaohua Yu, who organized, compiled, summarized, revised, reviewed, and checked this work. The academician Hequan Wu directed the final examination and checks. Shumin Cao conducted extensive preparatory work in the early stages of the project establishment, arrangement, and layout. The report was led by the Wuhan Research Institute of Posts and Telecommunications (FiberHome Technologies Group, Ming Cai, Wei He, Liang Chen, and Xinquan Zhang) and the China Academy of Information and Communications Technology (Xiaohui Yu, Zhiqin Wang, Chunzheng Lu, Jianbo Hu, Jiguang Cao, Xiayu Li, Shaohui Li, Shuning Pang, Song Luo, and Dong Min). China Telecom (Huiling Zhao, etc.), China Unicom (Zhongping Zhang, etc.), Tsinghua University (Jing Wang, Linling Kuang, Dan Li, Yong Cui, etc.), ZTE (Xiang Wang, etc.), Huawei (Yuedong Guo, Wenyang Lei, etc.), Shanghai Jiao Tong University (Wenjun Zhang, etc.), China Electronics Technology Group Corporation (Chunting Wang, etc.), Institute of Computing Technology Chinese Academy of Sciences and Beijing University of Posts and Telecommunications (Ping Zhang, etc.), Da Tang Telecom (Shanzhi Chen, etc.), Southeast University (Xiaohu You), Information Engineering University of PLA, and other units have made numerous contributions. We listened to the viewpoints of Shangyi Chen (Baidu), Jianming Zhou (China Mobile), the academician Jianping Wu (Tsinghua University), and Professor Xiaohu You (Southeast University). At the high-end forum of network and information held in Wuhan from November 3 to 4, 2016, Shaohua Yu listened to the viewpoints of the academicians Zuoning Chen, Zisen Zhao, Deren Li, Hequan Wu, Jiangxing Wu, Yunjie Liu, Yue Hao, You He, Jianping Wu, Chun Chen, Bangkui Fan, Xiangke Liao, Ming Liu, and Yongliang Wang. From December 2016 to January 2017, President Zhaoxiong Chen of the China Institute of Communications provided very good constructive opinions on the difficulties of industry integration, technology integration, and development after he consulted the relevant departments and bureaus of the Ministry of Industry and Information Technology. In

February 2017, the academician Yuemin Li offered two corrections and guidance on cyber security. In February 2017, the academician Jing Chen presented eight guidelines. The academicians Zuoning Chen, Xicheng Lu, Jiangxing Wu, and Manqing Wu of the Department Information and Electronic Engineering; Yaohui An; Guimei Fan; Da An; and Jiguang Cao have performed numerous work in this field. The second draft was supplemented by Shaohua Yu, who took the lead, organized, compiled, summarized, revised, reviewed, and checked this work. At the beginning of May 2018, he listened to the opinions of four academicians, Zuoning Chen, Xicheng Lu, Hequan Wu, and Jiangxing Wu. Shaohua Yu reported the core content of this topic network at the Second Global Future Network Development Summit held in Jiangning, Nanjing, from May 11 to 12, 2018, which had more than 300 participants. Wei He, Liang Chen, Xinquan Zhang, Meimei Dang, and Shaohui Li proposed numerous revisions to the second draft. In July 2018, the academicians Hequan Wu and Jiangxing Wu presented a revised proposal. From July 12 to 13, 2018, the academician Jiangxing Wu proposed a series of comments and suggestions on the compilation of 13 subfields. The theme of this edition has been revised and supplemented based on the 2017 edition. Several topics, such as integrated circuits, in-depth learning, IPv6, industrial internet, future networks, and information optoelectronics, have been added to this edition as a single-line brochure. I express my deep respect and heartfelt thanks to all comrades who participated in the compilation of the present version of this work.

Contents

List of Series Contributors

The guidance group and working group of *Research on the Development of Electronic Information Engineering Technology in China* series are shown as below:

Guidance Group

Leader: Zuoning Chen, Xichen Lu.

Member (In alphabetical order):

Aiguo Fei, Baoyan Duan, Binxing Fang, Bohu Li, Changxiang Shen, Cheng Wu, Chengjun Wang, Chun Chen, Desen Yang, Dianyuan Fang, Endong Wang, Guangjun Zhang, Guangnan Ni, Guofan Jin, Guojie Li, Hao Dai, Hequan Wu, Huilin Jiang, Huixing Gong, Jiangxing Wu, Jianping Wu, Jiaxiong Fang, Jie Chen, Jiubin Tan, Jun Lu, Lianghui Chen, Manqing Wu, Qinping Zhao, Qionghai Dai, Shanghe Liu, Shaohua Yu, Tianchu Li, Tianran Wang, Tianyou Chai, Wen Gao, Wenhua Ding, Yu Wei, Yuanliang Ma, Yueguang Lv, Yueming Li, Zejin Liu, Zhijie Chen, Zhonghan Deng, Zhongqi Gao, Zishen Zhao, Zuyan Xu.

Working Group

Leader: Shaohua Yu, Jun Lu.

Deputy Leader: Da An, Meimei Dang, Shouren Xu.

Member (In alphabetical order):

Denian Shi, Dingyi Zhang, Fangfang Dai, Fei Dai, Fei Xing, Feng Zhou, Gang Qiao, Lan Zhou, Li Tao, Liang Chen, Lun Li, Mo Liu, Nan Meng, Peng Wang, Qiang Fu, Qingguo Wang, Rui Zhang, Shaohui Li, Wei He, Wei Xie, Xiangyang Ji, Xiaofeng Hu, Xingquan Zhang, Xiumei Shao, Yan Lu, Ying Wu, Yue Lu, Yunfeng Wei, Yuxiang Shu, Zheng Zheng, Zhigang Shang, Zhuang Liu.

The expert group and writing group of *Research on the Development of Electronic Information Engineering Technology in China* series are listed as below:

Expert group		
Full name	Work unit	Position/title
Shaohua Yu	China Information and Communication Technologies Group Co., Ltd.	Academician
Hequan Wu	Chinese Academy of Engineering	Academician

(continued)

Expert group

Full name	Work unit	Position/title
Shumin Cao	Beihang University/China Academy of Information and Communications Technology	Professor-level Senior Engineer
Xiaohui Yu	China Academy of Information and Communications Technology	Professor-level Senior Engineer
Zhiqin Wang	China Academy of Information and Communications Technology	Professor-level Senior Engineer
Chuncong Lu	Telecommunication Administration Bureau of Ministry of Industry and Information Technology of People's Republic of China	Professor-level Senior Engineer
Jianbo Hu	China Academy of Information and Communications Technology	Professor-level Senior Engineer
Denian Shi	China Academy of Information and Communications Technology	Professor-level Senior Engineer
Meimei Dang	China Academy of Information and Communications Technology	Senior Engineer
Huiling Zhao	China Telecom Group Co., Ltd.	Professor-level Senior Engineer
Jianming Zhou	China Mobile Communications Group Co., Ltd.	Professor-level Senior Engineer
Shangyi Chen	Baidu	Senior Engineer
Shanzhi Chen	China Information and Communication Technologies Group Co., Ltd.	Professor-level Senior Engineer
Yuedong Guo	Huawei Technology Co., Ltd.	expert
Wenyang Lei	Huawei Technology Co., Ltd.	expert
Xiang Wang	ZTE Communications Co., Ltd.	Senior Engineer
Linling Kuang	Tsinghua University	Professor
Dan Li	Tsinghua University	Professor
Yong Cui	Tsinghua University	Professor
Wenjun Zhang	Shanghai Jiao Tong University	Professor
Ping Zhang	Beijing University of Posts and Telecommunications	Professor
Xiaohu You	Southeast University	Professor
Shaoqian Li	University of Electronic Science and Technology of China	Professor
Chunting Wang	The 54th Research Institute of China Electronics Technology Group Corporation	Professor-level Senior Engineer

Writing group

Full name	Work unit	Position/title
Shaohua Yu	China Information and Communication Technologies Group Co., Ltd.	Academician

(continued)

Writing group		
Full name	Work unit	Position/title
Liang Chen	China Information and Communication Technologies Group Co., Ltd.	Senior Engineer
Wei He	China Information and Communication Technologies Group Co., Ltd.	Senior Engineer
Xinquan Zhang	China Information and Communication Technologies Group Co., Ltd.	Senior Engineer
Meimei Dang	China Academy of Information and Communications Technology	Senior Engineer
Jiguang Cao	China Academy of Information and Communications Technology	Senior Engineer
Shaohui Li	China Academy of Information and Communications Technology	Senior Engineer
Song Luo	China Academy of Information and Communications Technology	Senior Engineer
Junsen Lai	China Academy of Information and Communications Technology	Senior Engineer

About the Authors

Chinese Academy of Engineering (CAE) is China's foremost academic and advisory institution in engineering and technological sciences, which has been enrolled in the first batch of pilot national high-end think tanks. As a national institution, CAE's missions are to study major strategic issues in economic and social development as well as in engineering technology progress, and to build itself into an S&T think tank having significant influences on decision-making of national strategic issues. In today's world, the wave of information technologies featured by digitalization, networking, and intelligence is gaining momentum. Information technologies are experiencing rapid changes with each passing day and are fully applied in production and life, bringing about profound changes in global economic, political, and security landscapes. Among diverse information technologies, electronic information engineering technology is one of the most innovative and widely used technologies and plays its greatest role in driving the development of other S&T fields. In order to better carry out strategic studies on electronic information engineering technology, promote innovation in relevant systems and mechanisms, and integrate superior resources, Center for Electronics and Information Studies (hereinafter referred to the "Center") was established in November 2015 by CAE in collaboration with Cyberspace Administration of China (CAC), the Ministry of Industry and Information Technology (MIIT), and China Electronics Technology Group Corporation (CETC).

The Center pursues high-level, open, and prospective development and is committed to conducting theoretical and application-oriented researches on crosscutting, overarching, and strategically important hot topics concerning electronic information engineering technologies and providing consultancy services for policymaking by brainstorming ideas from CAE academicians and experts and scholars from national ministries and commissions, businesses, public institutions, universities, and research institutions. The Center's mission is to build a top-notch strategic think tank that provides scientific, forward-looking, and timely advice for national policymaking in terms of electronic information engineering technology.

The main authors of *Network and Communication* are Shaohua Yu, Liang Chen, Wei He, Meimei Dang, and Shaohui Li.

Dr. Shaohua Yu, Academician of Chinese Academy of Engineering (CAE), information and communication network technology expert. He is the chief engineer of China Information and Communication Technologies Group Co., Ltd., the chief engineer of China information and Communication Technology Group Co., Ltd., the director of State Key Laboratory of Optical Communication Technologies and Network, the vice president of China Institute of Communications, the member of national 863 Program Network and communication subject expert group, the member of cyber power strategy research advisory group, and the national integrated circuit industry development advisory committee member. He has been engaged in the research of optical fiber communication and network technology for a long time, presided over and completed more than ten national projects such as 973 and 863, all of which have achieved transformation of achievements and a large number of applications. It is one of the pioneers of the integration of SDH (Synchronous Digital Hierarchy) and Internet (including Ethernet).

Chapter 1
Introduction

Currently, the entire world is participating the process of digitalizing, networking and intellectualizing the deep integration of the network information world with the natural world and human society. This process will lead to a significant improvement of the quality of productivity and introduce new methods of production and lifestyles. This advancement will have a profound and significant impact, similar to the agricultural and industrial civilizations experienced in human history. This process is profoundly changing competition throughout the world. Security, economics, society, military and culture trends generate new opportunities for national development, new spaces for people's lives, new fields of social governance, and new momentum for industrial upgrading and international competition. Over the next 20 years, the development direction of network communication technology will be human-network-thing three-domain interconnection and its systematic integration with various industries and regions. This direction will be realized by digitalization, networking and intellectualization, lead to the connection, extension and penetration of the natural world and human society, and the network space will be continuously enriched and expanded. With the acceleration of the deployment of the fifth Generation Mobile Communication System(5G), the number of network connections has increased from 10 to 100 billion and one trillion, from the interconnection of human-network two-domain to human-network-thing three-domain and multiple interconnections, and from plane interconnections on the ground to three-dimensional interconnections in space and interstellar interconnections (the information network, which is equivalent to adding a nervous layer to the natural world and human society; its essence is deep informationization). The network has the ability to perceive, transmit, store and process beyond imagination. Combined with traditional industries and regions, this network will create infinite possibilities. Accelerating the construction of "cyber power" in China is a strategic choice to seize the great opportunities of the information revolution and satisfies the urgent need to build a strong country in regard to science and technology, build cyber power and reshape its international competitiveness.

© China Science Publishing & Media Ltd (Science Press) 2020
Center for Electronics and Information Studies, Chinese Academy of Engineering,
Network and Communication, https://doi.org/10.1007/978-981-15-4596-2_1

The total amount of network data has rapidly increased from TB and PB levels to EB, ZB, YB and BB levels (1 PB equals 1000 T, which is the equivalent of approximately 4.4 U.S. National Library data volumes several years ago), and video occupies the majority of network traffic. Considering global industrial systems as an example, their systematic integration with the intelligent cloud, advanced computing, analysis, perception technology and the Internet has caused the emergence of an Industrial Internet system, which forms intelligent equipment, intelligent systems and intelligent decision-making and promotes the transformation and upgrading of traditional manufacturing to intelligent manufacturing. The objective world that is primarily depicted by analog quantities will gradually undergo complete digitization, which will be an important change in human beings for hundreds of years. The Internet of Things will be the engine of emerging industries following the mobile Internet. Gartner predicts that global Internet of Things equipment will reach 26 billion units in 2020, with a market size of $1.9 trillion. McKinsey predicts that the impact of the Internet of Things on the economy will reach 27,000–62,000 billion dollars by 2025. 5G wireless full coverage and the Gigabit access network will be accelerated and connect more than 500 cities in China, the Belt and Road countries in the world. Over the next 20 years, China's information economy added value, which considers the Internet an important carrier, is expected to exceed half of China's annual GDP.

The interconnection of all things enables network communication to affect production and lifestyles of broad masses of people, and network communication is everywhere. The social significance of the network is that it penetrates the limits of a human's physical and mental abilities, space and time and overcomes the limitations of a large number of human facilities and resources, such as libraries and encyclopedias, universities, hospitals, cinemas, factories, shops and post offices etc. The network is more important than any major invention in human history, such as the automobile, airplane, rocket, high-speed railway, and electric power. Traditional industries have begun to undergo subversive innovation, and changes in science and technology are earth-shaking. Human beings have also been inadvertently changed, and important issues, such as birth, age, illness and death, have been included in the scope of technical solutions. Human society and the natural world are constantly digitized, networked and intellectualized to generate new productivity, constantly change old patterns, and create new needs and models, which will produce a comprehensive cross-cutting frontier. The industrial form has changed from tangible assets to tangible and intangible assets, from heavy assets to light assets, from atomic management to bit management, from manual labor to intellectual labor, and from "ground, sea, air and space" to "ground, sea, air and space + cyberspace". More than 90 countries around the world have formulated Cyberspace Security strategies. Since the Prism Gate incident, the Internet has become a place where a country will live and die in the future. The Internet is related to the rise and fall of a country's destiny and changes. The American Futurist Alvin Toffler once predicted that whoever has information and controls the Internet will control the whole world. Accelerating the construction of cyber power in a country is a strategic choice to seize the great opportunities of the information revolution, satisfy the urgent need to build a strong scientific and technological country and reshape international competitiveness and the international trend.

Chapter 2
Global Development Trend

Network communication is at the peak of innovation-intensive applications, new business models and artificial intelligence. This communication has undergone accelerated integration into manufacturing, transportation, education, media, energy, materials and other industries, as well as 5G. Accelerating the layout, the network is promoting the digital, networked and intelligent transformation of global industry and leading to the constant production of new products, models and formats. The Internet has become an essential infrastructure for human production and life and an important strategic facility that affects the total situation of a country. The integration and penetration of cyber threats and traditional threats profoundly affect national security. Cyberspace is an important variable that affects politics, economics, society, culture and national security. Cyberspace has become the fifth territory of a country after ground, sea, air and space and is a strategic place to win a long-term competitive advantage for a country.

By the end of 2018, the Internet was developed in more than 220 countries and regions, with more than 3.896 billion Internet users (51.2 households per 100 people), 8.16 billion mobile phone users (107 households per 100 people), 5.286 billion mobile broadband users (69.3 households per 100 people), 942 million fixed telephone users (12.4 households per 100 people), and 1.075 billion fixed broadband users (14.1 households per 100 people) (cited from ITU). Every 10% increase in broadband population penetration can boost GDP by 1.38% (cited from World Bank data). From November 2018, the penetration rate of IPv6 users in the United States has reached 34.1%, and that in Japan has reached 26.3%. Belgium, India, the United States, Germany, Greece, Switzerland, Uruguay, Luxembourg, the United Kingdom and Japan are the top 10 countries in regard to the proportion of IPv6 users. The number of networking devices per capita in the world increased from 0.08 in 2003 to 1.84 in 2010 to 3.47 in 2015 and may increase to 6.58 after 2020. Human beings had accumulated 12 EB data throughout history before the commercialization of computers and reached 180 EB in 2006 and more than 1600 EB from 2006-2011.

© China Science Publishing & Media Ltd (Science Press) 2020
Center for Electronics and Information Studies, Chinese Academy of Engineering,
Network and Communication, https://doi.org/10.1007/978-981-15-4596-2_2

Researchers estimate that this number will quadruple every 3 years to 8 ZB in 2015 (cited from the Fourth Industrial Revolution) and 108 ZB or more by the end of 2020.

2.1 Frontier Innovation

The Internet remains the main framework that will support the global information transmission infrastructure over the next few years. P-bit level transmission, E-bit level switching, Gigabit access, 5G full coverage and the industrial Internet will gradually become reality for realizing the effective transmission of information in space and time. The total trend of development in the field of network communication presents 10 technical characteristics:[2]

First, network characteristics. The Internet is a great open tool and a symbol of science and technology in the twenty-first century. The Internet has gradually become a "scene" that is active anytime and anywhere. Similar to sunshine, air, water and electricity, the Internet has become a universal "existence". The Internet reflects the interconnection of the human-network-thing, big integration, big connection, big data, and new intelligence and has gradually penetrated the whole natural world and all of human society, which has led to a major change in human history.

Second, architecture characteristics, which reflect the four major changes of 5G network architecture, cloud network integration and cloud-side collaboration, namely, from a complex closed system to an open-source new SDN/NFV cloud network integration architecture, from an administrative management system and traditional networking thinking to internet thinking, and from passive adaptation to active, rapid and flexible response, to form the elements of the architecture unit. The source has changed from the traditional business relationship to the construction of a new ecosystem in the industrial chain. The Internet will remain the dominant architecture, and its global status, with the greatest impact, the broadest coverage and the largest number of users, will not change.

Third, connection characteristics. Metcalfe's law indicates that the value of the network is proportional to the square of the number of nodes connected to the network. The greater the number of nodes, the higher the value. The value and impact of 1.4 billion interconnected mobile phones substantially outweighs that of equivalent supercomputers, and the impact of connectivity is subversive, which changes the business model of many traditional industries. An Internet connection can constantly penetrate distance and time constraints, hearing constraints, vision constraints, ability constraints, knowledge constraints, and mental constraints and extend to all industries and regions, which reflects the characteristics of wide bandwidth, wide coverage, high throughput, and green energy saving.

Fourth, spatial characteristics, which show that network communication can penetrate the limitations of space and location and the limitations of a micro-world and biological environment, such as the limitations of high-temperature and low-temperature environments as well as high pressure and low pressure, the limitations of the seabed and the interior of the earth, and other limitations that indicate that an environment is not suitable for human beings. The network interconnection extends from a two-dimensional interconnection on the ground to a three-dimensional interconnection on ground, sea and air; a micro-world interconnection; and an outer space and interstellar interconnection. A connection extends to all physical dimensions of the earth.

Fifth, the software characteristics of the communication system, which enable it to continuously increase the flexibility and functionality of the system, upgrade and iterate quickly on demand, and have scalability, timeliness, reusability and reconfigurability. The goal is to define everything in the software and enable the utmost flexibility. However, the large-scale software in network communication system devices may be one of the most complex products that human beings can produce. For a software system with a large number of possibilities, testing all possibilities is difficult because it is a discontinuous system.

Sixth, data characteristics. The total amount of network data will rapidly increase from PB, EB and ZB levels to ZB, YB and BB levels, and the network traffic, storage capacity and processing speed will exponentially increase. These data reflect the big data presented in the network in recent years and data from focusing on causality to pay more attention to data relevance, timeliness and personalization. The collection, storage, management and analysis of data have become the focus of network information technology research.

Seven, bandwidth characteristics. According to Cooper's law, the amount of information transmitted in a given wireless spectrum doubles every two and a half years. Bit-per-second traffic on the Internet doubles every 16 months. From 1982 to 2018, single-wavelength commercial systems increased by 50,000 times. If we take into account WDM, the bandwidth will increase by 1.25 million times. The optical fiber transmission bandwidth of a backbone network doubles every 9–12 months. Compared with 30 years ago, the wireless gap rate of a 5G commercial system has increased by 100,000 times. The connection bandwidth will show development trend of Gbps - > Tbps - > Pbps - > Ebps - > Zbps, and the connection bandwidth will continue to grow exponentially.

Eighth, virtual characteristics, which primarily refer to Cyberspace and reflect the characteristics of no substance, no entity and no boundaries. Cyberspace has become a new field and space for competition and games among countries. Network (virtual) space has become the fifth territory and strategic space of human beings after ground, sea, air and space. Network rights are as important as sea rights, air rights and sky rights. Network space has become the focus of competition among countries. Many important functions of nature and human society are constantly shifting to virtual space to achieve nonmaterialization or dematerialization. The larger the number of Cyberspace resources that are employed, the more abundant the "site" that is developed.

Ninth, digital characteristics. The physical world, which is primarily depicted by analog quantities, will gradually undergo complete digitalization, which is the most basic part of the information society (similar to "measurement and its unification" more than 2000 years ago). These characteristics comprise a major transformation of the natural world and human society in parallel with the network; they extend the network resources and computing resources to any part of the world via analog-digital and digital-analog conversion and integrate the digital and analog worlds. From the beginning of digitalization (followed by networking and intellectualization) to the advanced stage of intellectualization, all traditional things will be redefined.

Tenth, micro-characteristics. The components of network communication are generally developing towards micro-miniaturization, increased integration, more functions, less energy consumption, less weight, smaller size and lower price. Over the past 50 years, Moore's Law has promoted the sustainable development of the integrated circuit industry. Currently, the planar layout of miniaturized transistors is approaching their physical and technological limits. The trend of Moore's Law will decelerate or change its orbit, and new methods and means are urgently needed to achieve breakthroughs.

After the several decades of exponential growth in the network bandwidth, the transistor number on chips and software code number, network communication technologies (such as the network bandwidth and transistor number on chips) will face seven "technology walls" around the next 20 years, including technology wall of the mobile communication, IP data communication, optical fiber communication, integrated circuit, software complexity, big data processing and energy consumption. These seven "technology walls" are relatively independent, interactive and self-contained. A major breakthrough in principle is urgently needed to support a large leap of network communication technology. These seven "technology walls" are described as follows:[3]

First, the mobile communication technology wall (first W), which lacks revolutionary breakthroughs in wireless communication theory over the past two decades, has approached the bottleneck of the further development of mobile communication technology. Against the background of the exhaustion of radio spectrum resources and the urgent need for a further substantial improvement in network bandwidth, we expect new means, theories and technological innovations to ensure the next leap-forward in the development of mobile communication technology.

The second, the IP data communication technology wall (second W). Since the establishment of the core technology system of TCP/IPv4, no essential breakthrough has occurred in basic technology. Internet IPv4 addresses are almost exhausted and the TCP/IPv4 architecture is facing the challenges of scalability, quality of service assurance, security, manageability and ability to carry hundreds of billions or trillions of interconnections and the huge traffic of the Internet. Important security events traceability is a difficult problem to overcome, and it is urgent to speed up research of the new generation of Internet technology.

The third, the optical fiber communication technology wall (third W), which has the goal of doubling the transmission bandwidth every 9–12 months and improving optical fiber communication efforts over the next 50 years. Over the past 10 years, the growth of optical fiber communication capacity has lagged far behind the demand of Internet traffic growth. Further improvement of its transmission capacity is limited by optoelectronic devices and optical fiber nonlinearity. A network transmission capacity crisis is expected in the next 20 years, and optical fiber communication urgently requires a breakthrough and subversive innovation.

The fourth, the integrated circuit technology wall (fourth W). Moore's law has been driving the continuous development of integrated circuits over the past 50 years. Currently, the planar layout of miniaturized transistors is approaching its physical and technological limits, and the trend of Moore's law will decelerate or shift (for example, memory chips can evolve in a three-dimensional direction, reduce costs and increase encryption). Therefore, new methods are urgently needed to achieve breakthroughs. However, the trends of increasing integration and speed, miniaturization, cost reduction and power consumption of integrated circuits will not change.

The fifth, the software complexity technology wall (fifth W). Currently, the number of lines of code in a large-scale network communication software system has exceeded 10 million lines. How much upgrade space is available? Assuming that the length of code will continue to double every 2 years, software systems will be able to accommodate 20 billion lines of code in 2040, which is 2000 times the current number. Over the next 20 years, it will be very difficult to maintain the original code scale growth rate (according to the original software programming method) or slow or change the growth mode, which urgently requires new and significant theoretical innovation.

The sixth, the big data processing technology wall (sixth W). Big data generally have the characteristics of volume, variety, velocity and value. People pay attention to causality instead of relevance, timeliness and personality. With the rapid leap of the total amount of network data from PB, EB and ZB to ZB, YB and BB, we have been unable to address big data with traditional ideas and methods. Massive information is challenging. New innovative breakthroughs and innovative means are needed. Our efforts are focused on improving processing capacity and solving interoperability.

The seventh, the energy consumption technology wall (seventh W). With an increase of the capacity, functions and power consumption of existing network equipment, it is difficult to sustain energy consumption by relying on traditional means, such as air cooling. A large data center consumes more power than small- and medium-sized countries. The energy consumption of end-to-end per-bit information transmission on the network will also encounter bottlenecks, and the trend of green energy conservation and the continuous reduction of power consumption will not change.

Currently, an important direction to promote the sustainable development of network communication over the next 20 years is the artificial intelligence of this super-large construction, which will further accelerate the evolution process of the ten technical characteristics. In the next 20 years, the artificial intelligence of network communication will integrate the cloud network, perception, big data and algorithms (don't need to face the pressure, motion, gravity, high and low temperature, underwater, and nanometer, which are robots face to, and also don't need the core cognitive functions of human beings), self-perception, self-adaptation, self-learning, self-execution, self-evolution, and network-based group intelligence responses. Using (network + AI) will become an important trend[4]. *Science* published the article The Power of Crowds in January 2016. This article proposes to solve the problem of rapid growth by combining group wisdom with machine performance. No convincing evidence that machines will never be able to match human intelligence (cited from "Intelligence is not artificial" by Piero Scaruffi) is available.

Experts predict that machine intelligence will compete with human intelligence by 2029; human will integrate with artificial intelligence by 2030; humans and machines will be deeply integrated by 2045; and the singularity is evident (which refers to the era when machines are smarter than human beings, quoted from Ray Kurzweil, Future of Artificial Intelligence). Human beings will be redefined by the two-way integration of "human omnipotence" and "machine intelligence" and have the ability to solve the most important personal problems, interpersonal problems, social problems, political problems, and economic problems. Human beings will achieve all-around development, will achieve their potential, and may "live a long life" or even "live forever". Five preconditions for this viewpoint exist. First, AI is producing remarkable results; second, technological progress is accelerating; third, human beings are helping machines create intelligence beyond that of human beings; fourth, human beings will benefit from intelligent machines that are smarter than us; and fifth, machines that pass the Turing test may be smarter than human beings.

However, complete machine intelligence does not conform to Godel's incomplete theorem (cited from Computer and Common Sense and Mind, Machine and Godel). Alonso Church extends Godel's incomplete theorem to the field of computers (cited from the Essence of Intelligence). Propositions exist in any formal system that cannot be proved or falsified by axioms and step rules (Roger Peng, Emperor's New Brain). Some experts also believe that it will be difficult for AI to reach the singularity. First, human beings will not cede power to smarter candidates; second, machines will not be subject to financial and resource constraints; third, smart robots will experience difficulty in managing the best of 7.4 billion people; fourth, even if machines can accurately reproduce human brain information, they will only clone but not transfer consciousness (quoted from "2040 Prophecy" Peter B. Scott Morgan). Doug Engelbart, the mouse inventor, had intended to use the term "increase wisdom" instead of "artificial intelligence" because he believed that machines would make our intelligence stronger.

AI is currently better at addressing closed and specific problems of certainty, primarily through machine learning (depth learning) and symbol representation and reasoning, such as Go, which is a typical problem of space expression and a search problem. Once computer intelligence is comparable to that of human beings, with its advantages of capacity, speed, memory, search and accuracy, levels beyond human intelligence may exist! If this day arrives, then the machine is likely to be invincible.

2.2 Technological Innovation

From the perspective of technology development, network communication has experienced four stages: analog, digital, internet and mobile internet. The development trend of network communication is 3 s-abc[5], that is, "3" refers to the three-billion-level human-network-thing interconnection (connecting the whole world, 5G full coverage and Gigabit access) and deep integration with various industries and technologies; "s" refers to a software-defined network, network function virtualization and open source (SDN/NFV/Open Source); "a" refers to the network interconnection and technology integration realized by artificial intelligence; "b" is the big data of the network; and "C" refers to the cloud of the network. Large bandwidth, wide coverage, high throughput, miniaturization and green energy-saving are the basic characteristics of the connection. All services are user-centered.

First, wireless mobile communication is trending towards "5G/6G". The development trend is 3X s-abc, where "3" refers to the interconnection of human-network-objects at the level of 300 billion and the deep integration of 5G with various industries; "X" refers to the optimization technology for the electromagnetic environment and application mode, including higher speed (greater than 1 Gbit/s), wider coverage, larger-scale connection and lower delay (millisecond level); "S" refers to the software-defined network, network function virtualization and open source (SDN/NFV/Open Source); "a" refers to the artificial intelligence of a wireless mobile network; "b" refers to the big data of a wireless mobile network; "c" refers to the cloud of a wireless mobile network; and all services are user-centered. Currently, a 5G/6G baseband chip, RF front-end receiving device, high-end processor chip and real-time embedded operating system are research and development difficulties. In addition to satisfying the 1000-fold expansion demand of mobile internet services over the past decade, high reliability and low latency communications, open network architecture reconstruction, flexible end-to-end slicing and multiform customized terminals, the application scope of 5G/6G has expanded from current human-network communication to the more extensive areas of human-network-thing three-domain collaborative communication, the ultra-dense Internet of things, vehicle network and industrial internet. Terahertz communications, large-scale antennas, spectrum sharing, wearable equipment networking, artificial intelligence, robots, driverless vehicles, smart cities, edge computing, digital identity and other technologies are emerging. Significant and irreplaceable killer applications are difficulties of 5G. 5G will lead to great changes in the entire telecommunication industry.

Second, the IP data network will develop toward "Ts-abc". "T-bps level or higher" is line speed forwarding direction, s is the software definition network, network function virtualization and network software open source (SDN/NFV/Open Source) direction; a is the data network artificial intelligence, b is big data and c is cloud evolution. Currently, the core devices, P-bit level switching chips and high-speed processor chips used in T-bit line cards are research and development difficulties. IPv4 will gradually transition to IPv6. The internet will extend to the field of manufacturing, and the industrial Internet will become increasingly popular. The IP data network will gradually transform to network resource virtualization, network function virtualization, network operation virtualization, network service openness, network resource openness, network technology openness and open source transformation. Network software, flexibility, and miniaturization will continuously promote the innovation of network architecture and a forwarding mechanism and promote network addressing and routing scalability, manageability and security. Credibility innovation promotes core chip and network operating system innovation. Currently, the traceability of important security incidents is a problem in network supervision. Research on IPv6 servers has been continuously strengthened, and the internet has entered the "post-IP" era, which is a general trend. Developing a new generation of Internet technology, penetrating the limitations of the TCP/IPv4 protocol; building a future network, and realizing intelligent, open and open source platforms; an adaptive security architecture; and an industrial Internet system are the directions of industry's efforts and promote the network to be open, intelligent, ubiquitous, integrated and integrated. Open source software, open source hardware, future networks, information traceability and mining, space interconnection, digital home, 4 K/8 K ultra-high definition video, VR/AR, 3D printing, digital identity, block chain, quantum computing, pervasive computing, edge computing, brain-like computing, brain-computer interface, cognitive computing, protocol reconstruction, DNA storage, information physics system (CPS), full-dimensional perception, 3D video, self-perception and self-immunity, end-to-end level protection for full-node, universal unlimited storage, and neural technology have separately emerged.

Third, optical fiber communication will develop towards IPs-abc, photoelectric integration ("I") and silicon-based integration, and high-speed, large-capacity and multidimensional parallel multiplexing of "P-bps". Optical fiber communication will develop towards a software-defined optical network, functional virtualization of optical network and open source of optical network software (SDN/NFV/Open Source), artificial intelligence of optical network ("a"), optical network transmission support, business big data ("b") and optical network cloud ("c") direction evolution. The development of integrated optical chips, photoelectric integrated chips, AD/DA and DSP for ultra-high speed, ultra-large capacity and ultra-long distance optical fiber transmission is difficult. In the world of optical connectivity, and multidimensional optical connectivity (long distance, metropolitan area, access, data center optical interconnection, short distance, inter-shelf, inter-board, inter-chip, and in-chip), "light" is widely used by ordinary people, such as for water and electricity, while miniaturization, universality, indispensability, inexpensive, anytime and anywhere and the social trends of optical communication. The share of optical interconnection will gradually exceed that of traditional electrical interconnection. The distance of optical interconnection exponentially decreases and extends from inter-frame to inter-board, inter-chip or even intra-chip, while the number of optical interconnections exponentially increase. Pbps-level ultrahigh speed backbone optical transmission, Tbps-level ultra-low-cost metropolitan optical transmission, diversified and wide-coverage low-cost ultra-wideband optical access, ultra-low delay, ultra-high precision time synchronous transmission, silicon-optical integration and optoelectronic integration, software-defined optical network and network virtualization, multidimensional multiplexing, all-optical network, ultra-low-loss optical fiber, ocean cable system, optical fiber sensing, ocean monitoring network, space optical nodes, deep-sea optical nodes and ground-sea-air-space optical networks are the foci of technological innovation. New technologies such as nanolasers, optical storage, atomic optical switches, memristors, quantum communications, nanotechnology, graphene and its CMOS, carbon nanotubes, micro-nanotechnology, optical field display, dynamic reconfigurable and on-chip network technology, CMOS high-speed interconnection with a data rate greater than 100G bit/s, and 3D integrated circuit design are coming to the front. New technologies are emerging. Optical fiber access will develop to 50G-EPON (802.3ca) and 1∗50/N∗50G PONs with a higher rate. With the development of network communication technology, especially cloud computing technology, big data, artificial intelligence and new data centers, optical fiber access will effectively promote the development of network communication to an ultra-wideband "cloud" network and accelerate the integration of IT and network communication technology.

A DRAM chip in can contain 16 billion transistors (entering the stage of mass production). Assuming that we continue to follow the trend of Moore's law, we will be able to integrate 4.8 billion transistors on a single chip in 2040, which is 30,000 times the current number and 30,000 times the current processing power.

2.3   Industrial Development

Generally, the development of the global telecommunication equipment industry has entered a mature period, and the growth of market scale is gradually decelerating. The network communication industry is facing the transformation of hundreds of billions of interconnected things, and network structure reconstruction is the next step, which will affect the mode of industrial development and the ecological environment. First, with open reconfiguration and open source software as the leading factor, the rate of development of the industrial layout of "standardized hardware + system architecture reconfiguration/network function virtualization" is increasing, the software and hardware architecture of the open source mode is gradually influencing the mainstream, and the network "customized service" mode will extend to network communication operation Second, the form of industrial competition is changing from product and engineering application competition to product customization and differentiated service experience competition that fully satisfies the need to continuously satisfy the potential and deep-seated needs of users. An open software and hardware platform ecosystem is forming in the field of network communication and will produce a variety of new production, service and business models.

In October 2018, the US President instructed his Department of Commerce to develop a long-term and comprehensive National Spectrum Strategy. In April 2019, the US announced plans to invest $275 billion to accelerate the rollout of 5G in the United States. The EU issued a 5G action plan in 2016 that will allocate a

3.4–3.8 MHz band and part of the 26 GHz band to 5G before 2020; each member country will choose at least one city to provide 5G services. Japan released 5G in 2016 and proposed commercial 5G during the Tokyo Olympic Games in 2020. South Korea proposes to demonstrate a 5G pilot during the 2008 Pingchang Olympic Games and start providing commercial 5G on April 3, 2019. Completion of full coverage is expected in 3 years. Russia expects to achieve 5G coverage in large cities by 2020 and to extensively employ 5G by 2024.

Because 5G is strongly industry-driven, in recent years, 5G has become the focus of fierce competition among global manufacturers. The formulation of 5G international standards has accelerated. The 5G NR Standalone Architecture(SA) standard was approved by 3GPP in June 2018, and industrial competition regarding 5G has re-entered the sprint stage. To solve the standardization problems of enhancing the performance of mobile broadband (eMBB) and the low latency and high reliability of Internet of Things, such as the industrial Internet, and vehicle networks, 3GPP has entered the second stage of 5G standard formulation. Three scenarios are supported: enhanced mobile broadband, mass machine communication (mMTC) and low delay and high reliability the Internet of Things. The 5G technology level and intellectual property rights of Chinese manufacturers are in the first echelon of the world; in the future, however, they still need to continue to develop in core chips, operating system software, ultra-intensive networking, industrial Internet, vehicle networking and other technologies. 5G has the characteristics of higher speed, wider coverage, larger scale connection, lower delay and higher reliability. In terms of enhancing mobile broadband, 5G can provide a 1 Gbit/s user experience rate and 10 Gbit/s peak rate, which is nearly 100 times higher than 4G. In terms of the number of user terminals, 5G supports millions of terminal connections per square kilometer and is 1000 times faster than 4G. In terms of network delay, 5G can provide a millisecond end-to-end delay for users, which is 10 times lower than 4G. Indicators are expected to be upgraded by an order of magnitude; for example, the American operator Verizon launched 5G mobile services on April 3, 2019, between Chicago and Minneapolis with a measured rate of 762 Mbit/s and a delay of 19 Ms.

Chapter 3
Current Development of China

In China, internet users, telephone users, mobile broadband users, fixed internet broadband access users and fixed telephone users account for 21.3%, 21.4%, 24.8%, 37.8% and 19.3% of the total users in the world, respectively. China's online retail trade volume ranks first in the world. In 2018, China's total telecommunications business reached 6555.6 billion RMB, which is an increase of 137.9% from that of the previous year. The telecom business revenue totaled 1301 billion RMB, which is an increase of 3.0% from the same period of the last year. In 2018, China's fixed telecommunications business revenue reached 387.6 billion RMB, which is an increase of 9.1% from the previous year's telecommunications business revenue of 29.8%, indicating an increase of 1.7 percentage points from that the previous year. China's mobile communications business revenue reached 913.4 billion RMB, which is an increase of 0.6% from the previous year's telecommunications business revenue of 70.2%. In 2018, China's voice business revenue reached 177.6 billion yuan, which was annual decrease of 25.7% year-on-year, and its share of telecommunications business revenue decreased to 13.7%, a reduction of 4.2 percentage points from that of the previous year. In 2018, China's fixed data and internet business revenue reached 207.2 billion RMB, which was an increase of 5.1% from that of the previous year, and its share in telecommunications business revenue increased to 15.9% from 15.6% of the last year. China's mobile data and Internet business revenue reached 605.7 billion yuan, which was an increase of 10.2% from that of the previous year, and its share in telecommunications business revenue increased from 43.5 to 46.6%. The IPTV business revenue increased by 19.4% compared with that the previous year; the Internet of Things business revenue increased by 72.9% compared with that of the previous year (cited from Ministry of Industry and Information Technology). In recent decades, China's wireless bandwidth has increased by 100,000 times; the optical fiber bandwidth has increased by 1000 times; the size of integrated circuits in China has decreased by 10,000 times; the performance has improved by 100,000 times; the cost has decreased by 10,000 times; and the amount of software code has increased by 100,000 times [1].

© China Science Publishing & Media Ltd (Science Press) 2020   13
Center for Electronics and Information Studies, Chinese Academy of Engineering,
Network and Communication, https://doi.org/10.1007/978-981-15-4596-2_3

3.1 Frontier Innovation

In recent years, China's network communication technology has accelerated the transformation from the follow-up mode to the innovation-driven mode, which highlights the theoretical basis of innovation, and key technology research constantly impacts the international advanced level. A series of outstanding achievements has been achieved. First, in the field of data communication, the latest progress in global future internet architecture research has been closely followed, and several future network architectures and system theories have been independently proposed. Including innovation on future network platforms, full-dimensional definable network architecture and baseline technology, a small-scale test and verification platform for future networks has been constructed. Second, basic research in the field of mobile communication and advanced global levels are interconnected, and important progress has been made in the research of 5G new network architecture, such as intensive networking, high-throughput collaborative networking and virtualization of the wireless access network. A batch of 5G wireless transmission technologies, such as large-scale antenna arrays, channel coding, and efficient cooperative transmission, has been produced. Third, in the field of optical fiber communication, the monopoly of intellectual property rights on basic original technologies in the United States, Japan and Europe has been gradually weakened, and important breakthroughs in ultra-high-speed, ultra-large-capacity and ultra-long-distance optical transmission experiments, which have approached and have reached the world advanced level several times, have been performed. Basic research and transmission of quantum key applications has emerged. Although the research and development of communication network equipment and system key technologies in China has entered the international advanced ranks, and distinct shortcomings exist in high-end core chips and devices, key materials and high-end process equipment, and operating systems, which are restricted by developed countries. Strategic advanced research, basic theory, and system and concept innovation remain relatively weak, and the situation that national network information security and information industry security are subject to human beings has not fundamentally changed.

3.2 Technological Innovation

The field of network communication has become a model of independent technological innovation in China and has achieved numerous innovative achievements with substantial influence. A breakthrough of technological development from following and running part to part and running side by side with a small number of leading countries has been realized, and China has become an important co-runner in the global technological direction. First, the innovative capability of data communication technology has been significantly enhanced, which has led to the completion of more than 100 international standards, such as IETF, ITU and ONF, and significant progress has been made in IPv4/IPv6 interoperability, cyber security, routing protocols, software defined network (SDN) and network function virtualization (NFV). Second, optical fiber communication technology has become one of the most advanced high-tech fields in China and the world in the form of the optical network. The share of network system equipment in China is close to half of that in the world, and the share of access network system equipment has reached 3/4 of that in the world, which has enabled the formulation of packet transport network (PTN) and 40G/100G PON ultra-wideband optical access standards and actively contributes to the research results of software-defined optical networks in major open source organizations. Third, the independent innovation ability and global influence of the mobile communication industry have been significantly enhanced, which has enabled the completion of more than one-third of LTE-related projects. The IMT-2020 (5G) Promotion Group was established in 2013 to support the research and development of a new generation of mobile communication technology and the formulation of international standards. The group has had an important role in the development of 5G requirements and network architecture of international organizations, such as 3GPP. Some technologies have been adopted, and the 5G technology R&D test was launched in 2016. The second phase of the test was completed in 2017. The third phase of the system verification test is currently underway, and the 5G pre-commercial deployment is planned for 2019. New directions, such as a 5G mobile communication network and Internet of Things, Next Generation Internet, T-bit ultra-high-speed optical network, mobile Internet, future network, and quantum communication network, are rapidly developing, and infrastructure construction, such as high-speed broadband, intelligent convergence and narrowband Internet of Things, is accelerating.

3.3   Industrial Development

In 2018, the added value of the electronic information manufacturing industry above the scale increased by 13.1%, which is a faster increase than that of all industries by 6.9 percentage points. December 2018 experienced 10.5% year-on-year growth. In 2018, the export delivery value of the electronic information manufacturing industry above the scale increased by 9.8%, and the growth rate fell by 4.4 percentage points compared with that in 2017. In December 2018, the growth rate was 2.0% [6]. China's network communication manufacturing industry has transitioned its development path from introduction and absorption to innovation and has basically formed a complete industrial chain and advanced and independent network communication industry system. The R&D capability of some system equipment products has reached or approaches the international leading level. The industry scale is larger than that of developed countries. The export share is the first in the world. The optical transmission system and optical access system are the strongest, followed by mobile communication, and the router switch (data communication equipment) is slightly weaker. High-end core devices and operating system software remain weak links, and most of the independent core technologies are at the low end of the value chain. First, in the field of data communication, we focus on producing high-end core routers and switches. 400G/1T core routers have reached the advanced level in rest of the world while laying out the next generation router architecture research and development. Second, in the field of optical fiber communication, we are actively developing 400G/1T and higher-speed ultra-high-speed optical fiber transmission equipment and large-capacity networking equipment, which have made remarkable progress in optical network intellectualization and flexibility. A new system of ultra-large capacity, ultra-low delay and flexible networking has been developed while upgrading 5G forward and return transmission. Third, in the field of mobile communications, LTE-A technology research and development and equipment industrialization have been accelerated due to the development and application of large-scale antenna arrays and apos. 4G products and large-scale antenna arrays and the test performance of 5G has been outstanding. Investments in 5G mobile communication core technology have been achieved. In new and commercial networks, the entire machine is strong and the core device is weak. Our enterprises rank first in the world in regard to technology, patents, standards and products of 5G equipment. However, the core technical capabilities of radio frequency devices, baseband processing chips, general digital chips, and general CPU and operating system need to be improved.

3.4   Opportunities and Challenges

In 2018, China's digital economy totaled 31 trillion RMB, which accounts for one-third of its GDP in this year. International 5G deployment has entered a critical period of commercial use. The 5G Standalone Architecture(SA) standard has been determined by 3GPP. The first-stage standard is being perfected, and the spectrum plan has been released. In 2016, the United States released a 5G high-frequency spectrum plan, and AT&T plans to launch 5G broadband services with SA in 2018. South Korea auctioned the 3.5 GHz and 28 GHz spectra in June 2018 and plans to launch 5G commercial services in March 2019. Japan's spectrum allocation plan will be completed by the end of 2018. Both Japan and Europe plan to start providing 5G commercial services by 2020. Operators in Europe, America, Japan and Korea are actively exploring businesses, such as artificial intelligence, big data, cloud computing, industrial 4.0, industrial Internet, high-definition video, vehicle networking, and virtual reality. With the acceleration of global information technology reform, the adjustment of the international industrial structure and the in-depth implementation of China's cyber power and manufacturing power, China's network communication industry has ushered in a new stage of rapid development with a series of conditions and opportunities to achieve leapfrog development [7]. First, the opportunities for a scientific and technological revolution and technological change, the interconnection of human-network-thing, the great integration, big connections, big data and new intelligence provide a rare historical opportunity for China to achieve leapfrog development. Second, due to the immense market potential, China is in the critical period of completing its 200-year goals, and the systematic integration of network communication technology with various industries and regions provides China's network with great potential. The network communication market has provided immense demand. Third, China has a strong industrial foundation that relies on the industrial foundation and resource accumulation in the field of Internet and network communication in China and is expected to lead to the further growth of the industry by virtue of the advantages of China's manufacturing development policy.

China's network communication industry is facing some problems and challenges; its technology shortcomings are distinct. Compared with developed countries, such as Europe, US, Japan and Korea, China has weak capabilities in regard to high-end core devices, process equipment, instrumentation, and basic software and is generally ranks in the second tier in the world. First, the external dependence on core technology and equipment is high, and the self-sufficiency of high-end core chips, process equipment and basic software needs to be urgently solved. Second, a new generation of information technology industry clusters with world influence has not formed, the ecology of independent technology industry needs to be improved, a single breakthrough has not yet formed group advantages, and the supporting capacity of industry needs to be strengthened. Third, basic research needs to be strengthened, and innovative, basic and pioneering technologies are in urgent need of major breakthroughs. Fourth, to implement the goals of national science and technology power, cyber power, manufacturing power, speed-up and fee-reduction, higher, stricter and more urgent requirements are proposed for network technology innovation. In addition, the global cyberspace confrontation has been upgraded in an all-around way due to the transformation of the cyber security strategy from active defense to global deterrence by individual powers. Cyberspace has evolved into a place of war, which will be another version of modern warfare, including conventional and unconventional warfare. The method of warfare has changed, and its strategic significance has exceeded that of nuclear weapons. The network battlefield has become increasingly localized and globalized. The boundaries between the cyber army and civilians have become increasingly blurred. The boundaries between cyber warfare and peace have also blurred. New deadly technologies (such as stealth UAV, military robots, automatic weapons, cyber bombs and cyber space warfare) are more easily accessible. Important national strategies and information facilities can easily become targets of a network intrusion and precise attack. To subdue a country, the use of a nuclear weapon is not necessary; cutting off its network, water source, power supply, or transportation for a few days is sufficient.

Chapter 4
Future Prospects in China

4.1 Frontier Innovation

Aimed at cyber power and manufacturing power, a batch of basic theories and frontier technologies are laid out to strengthen areas of weakness. First, the field of mobile communication boldly explores the breakthrough direction of the new generation of wireless communication theory, explores the revolutionary technology and network architecture in the "post 5G" era, and develops new spectrum resources and means, such as terahertz and visible light. Second, in the field of data communication, the innovation of future network technology integrates the existing resources of various scientific research projects and experimental networks, accelerates the establishment of national future network regulations, models the test bed and promotes the interconnection and interoperability with existing platforms abroad. Third, in the field of optical fiber communication, we will perform new future-oriented technology research, such as T/P bit-level transmission, silicon and III-V photoelectric-integrated chips, 5G optical communication forward-return-transfer, ultra-dense optical access, SDN/NFV, ultra-low loss optical fibers and large effective area optical fibers, and explore new materials and processes for the preparation of optical fibers. Research has begun on the sixth generation mobile communication (6G), P-bit optical transmission system; T-bit ultra-long-span optical transmission; new optical access with wide coverage and large capacity; physical layer technology of large-scale wireless communication; new antenna and radio frequency technology for large-scale wireless communication for base station; terahertz wireless communication; and large-scale secure and trusted addressing routing. We will strive to change the situation that our core technology in the field of network communication is subject to human beings by 2025 and strive to make our network communication successfully rank among the world's best by 2030 to help our country enter the forefront of the world's most innovative countries.

© China Science Publishing & Media Ltd (Science Press) 2020   19
Center for Electronics and Information Studies, Chinese Academy of Engineering,
Network and Communication, https://doi.org/10.1007/978-981-15-4596-2_4

4.2   Technological Innovation

The overall basic research, technology research, equipment development, platform construction and application demonstration focus on optical networks, mobile networks, IP data networks, satellite communication networks, mobile Internet and the Internet of Things. It is important to research and establish a new generation of international advanced, autonomous and controllable network technology systems and standards and develop 5G follow-up technology (6G), ultra-high speed intelligent optical transmission, and high-speed packet transmission. With key technologies such as access, SDN/NFV, new routing switching and medium-low orbit broadband communication satellites, we have developed a series of key equipment and systems with the new generation network architecture, realized key breakthroughs and industrialization of related network chips and protocol software, performed a network technology application demonstration and experimental verification and promoted the total breakthrough of network communication technology in China. By 2020, China's IP data network field will become an important participant in global technology and industry as well as be a leading force in global technology research and development and international standards formulation in the field of mobile communications. The optical fiber communication industry is in a parallel phase with that of developed countries and leads the development direction of international frontier industry technology to a small degree; the broadband satellite communication field will focus on inter-satellite and satellite-to-earth. Breakthroughs have been achieved to narrow the gap with the advanced international level by a large margin.

4.3   Industrial Development

The goal of industrial development is to foster a new generation of information and communication technology industry ecology, to achieve leapfrog development of network communication industry, and to help China achieve cyber power. First, the basic general technology, asymmetric killer technology and frontier subversive technology in the core areas of optical fiber communication, IP data communication and mobile communication are separately sorted, and then, targeted development strategies are formulated to complement areas of weakness, for example, core chips, high-end equipment and basic software strengthening. Second, basic research, technological innovation and engineering applications are efficiently connected to upgrade the network infrastructure. Evolution and Internet application innovation are the leading factors for promoting network technology innovation and industrial development. Third, we should strengthen the vertical and horizontal cooperation of the whole network communication industry, establish an open industrial ecological system, enhance global power in both the open source software community and the international standard-setting, and penetrate the key technologies of the industry in common. Fourth, we should guide innovation carriers from a single enterprise to multiowners across fields. The transformation of the total synergistic innovation will promote the transformation of the innovation process from a linear chain to a synergetic parallel chain and encourage the transformation of the innovation mode from a single technological innovation to a combination of technological innovations and business model innovations to cultivate a number of internationally influential industrial clusters. Fifth, we should further improve the level of integration of industrialization, strengthen the integration and innovation of information and communication technology and traditional industry technology, and constantly promote the emergence of new technologies, business models and new directions of industry. In May 2018, according to "Internet Queen" statistics, among the world's top 20 Internet companies in regard to valuation or market value, China had 9 companies and the United States had 11 companies. In 2013, China had only two companies, while the United States had nine companies.

Chapter 5
Hot Spots in China

From the perspective of technology and application, the field of network communication is divided into four basic technical categories, mobile communication, data communication, optical fiber communication and quantum communication, and three application service categories, mobile internet, internet of things and edge computing.

5.1 Mobile Communication

In 2018, the total number of telephone users in China reached 1.75 billion (21.4% of the world), which is a net increase of 137 million and an increase of 8.5% from that of the previous year. The total number of mobile phone subscribers reached 1.57 billion (19.2% of the world), which is a net increase of 149 million, and the penetration rate of mobile phone subscribers reached 112.2 per 100 people, an increase of 10.2 per 100 people over that of the previous year. The total number of mobile broadband subscribers (i.e., 4G and 3G subscribers) reached 1.31 billion (24.8% globally), accounting for 83.4% of mobile phone subscribers, a net increase of 174 million over the entire year. The total number of 4G users reached 1.17 billion, with a net increase of 169 million over the entire year (citing data from the Ministry of Industry and Information Technology).

5.1.1 Technical Progress

Mobile communication has experienced four stages of evolution: analog, digital, multimedia and mobile ultra-wideband. China's wireless mobile communication technology has experienced four stages of evolution: 1G blank, 2G follow-up, 3G breakthrough, 4G parallel (synchronized with advanced foreign level), and 5G, for

© China Science Publishing & Media Ltd (Science Press) 2020   23
Center for Electronics and Information Studies, Chinese Academy of Engineering,
Network and Communication, https://doi.org/10.1007/978-981-15-4596-2_5

which the goal is to lead. 1G to 3G is mainly for personal communication. 4G has just expanded to the Internet of Things, and 5G deploys the IoT in an all-round way. 4G is also oriented to the industry and social management, such as the Internet of Vehicles. In 2017, more than 7.4 billion mobile subscribers existed worldwide, and the scale of the 4G network steadily increased. More than 400 LTE commercial networks have been deployed in more than 100 countries and regions [8]. With the large-scale commercial phase of 4G, the fifth generation mobile communication (5G) facing 2020 and the future has become a global industry hotspot. The Internet is expected to extend from consumer to industrial and other fields. 5G has the characteristics of a higher speed (more than 1 Gbit/s), larger capacity, lower delay (1 ms), higher reliability and lower power consumption. 5G will become the basic communication mode that connects people with people, people with things, things with things in 2020 and the future. The vision of 5G and its key capability requirements have been clear. 5G supports three scenarios, enhanced mobile broadband (eMBB), ultra-reliable ultra-low latency communication (URLLC) and massive machine type of communication (mMTC), and determines the user experience rate, latency, connection density, mobility, peak rate and other indicators. The evolution of the 5G core network is likely to adopt a unified internet/mobile network architecture (including naming, authorization and mobility management), distributed control, plug-and-play support for base stations, access point AP, and flat networks. 5G will be commercially available around 2020.

At the end of 2017, 3GPP issued the R15 Non-Standalone Architecture (NSA) protocol, which focuses on supporting the enhancement of mobile broadband services. The 5G base station is connected to the 4G base station or the 4G core network. After users access the network through the 4G base station, new 5G and 4G air ports jointly provide data services, and 4G is responsible for mobility management and other control functions. In June 2018, 3GPP released the 5G international standard to support SA to enhance mobile broadband and basic services with low delay and high reliability. Based on the full service architecture of a 5G core network, a 5G base station can directly connect a 5G core network and provide new applications, such as network slicing and edge computing. The third sub-stage R15 standard released in March 2019 completes additional networking architectures to support 4G base station access to the 5G core network. In addition, 3GPP will release the R16 standard in March 2020. Based on the R15 standard, the R16 standard will further enhance the ability and efficiency of the network to support mobile broadband while expanding support for more Internet of Things scenarios.

In terms of the frequency spectrum, in November 2017, China issued a medium frequency plan for 5G, which defined a total of 500 MHz for 5G from 3.3–3.6 GHz and 4.8–5.0 GHz. By the end of 2018, China completed its 5G medium-frequency spectrum planning and actively performed millimeter-wave spectrum research. In terms of technology research and development tests, according to the general plan of the 5G test, since the start of 5G technology research and development test in early 2016, the first and second phases of the test have been completed, and the third phase of the test is currently being performed. By the end of 2019, 5G system equipment will satisfy commercial needs. In terms of 5G applications, China held the 5G application contest, which solicited 5G applications and promoted the integration of 5G and vertical industries. Many 5G technology projects in China have successively entered the international core standards, and the speed, quantity and quality of their advancements rank among the top in the world. 3GPP and ITU plan to complete the 5G international standards by the end of 2019 and 2020, respectively. China's first 5G mobile phone was officially launched in the second half of 2019, and the iconic killer application is a 5G problem that must be overcome.

By freezing the 5G R15 version standard in June 2018 and determining the domestic frequency band, domestic and foreign manufacturers have accelerated the development and launch of predictable commercial products. For the macro station and indoor micro station, 2.6/3.5 GHz has reached the pre-commercial level, and the 4.9 GHz band was used as the prototype. Domestic manufacturers have launched a variety of products, such as the macro station, small station and micro station, to support indoor and outdoor scenarios. Throughout the world, 1.2 Gbit/s second-generation Gigabit LTE modem chips have been released. RF front-end and receiver chips have been developed to support the 16QAM, QPSK 20 MHz bandwidth. Commercial GPUs have been developed to 2peta FLOPS, with 512 GB of display memory. For the core network, the NSA is commercially available. Based on the core network scheme of SA, due to the introduction of new technologies such as SDN and NFV, 5G is close to commercial use. A test network has to be set up. Complete commercial use requires further efforts. 5G mobile phones are commercially available but need to be strengthened, especially in regard to their chips. The majority of the mobile phone manufacturers in China announced the launch of 5G mobile phones in 2019. From the "neck-jamming incident", we observe that the key core technologies need to be complemented, and the global competition for 5G core technologies is becoming increasingly fierce.

China's satellite communication technology continues to maintain a steady development trend. Small satellite technology and high throughput (HTS) in-orbit satellite technology have rapidly developed. High-orbit and low-orbit satellite Internet, inter-satellite and satellite-to-ground laser communication have attracted the attention of the industry.

5.1.2 Trend Prediction

Wireless mobile communication will develop towards "5G/6G". Open 5G/6G system architecture will be adopted, and 3Xs-abc will be the main development trend. With OFDM and MIMO as the core technologies, the development of 4G-LTE mobile communication is on the rise and is gradually expanding as a basic technology. China's 4G users remain in a stable period of development. LTE-V, LTE-U, LTE-D and other technical standards for vehicle networking, the Internet of Things and terminal through applications are gradually maturing. Narrow-band LTE (NB-IoT and eMTC) and the industrial Internet for small data, big connection and wide coverage of Internet of Things applications have attracted extensive attention from the industry and are expected to become a wide-ranging IoT operation infrastructure. Thus, a new direction emerges for the development of IoT applications based on a public network.

5G will be commercially available around 2019. The availability of 5G will expand from the current human-network communication to the human-network-thing three-domain interconnection, the ultra-dense interconnections of the Internet of Things, the vehicle network and the industrial Internet. Network end-to-end slicing technology has become a research hotspot in the industry to satisfy the diverse application requirements of the mobile Internet and Internet of Things. Wi-Fi, as the representative of broadband wireless access technology, is also developing towards wider bandwidths, higher speeds and more services. Many new applications have been developed, including Internet of Things applications using the 802.11ah protocol, vehicle network applications using the 802.11p protocol, and low latency and large bandwidth applications using the 802.11ad protocol.

Nearly 80% of the land area and 95% of the sea area for communication coverage are blind or weak areas in China. The space-based information network based on near-earth and high-orbit satellites has the characteristics of high, far and wide coverage compared with the ground information network. The network has distinct advantages in realizing communication in remote areas, which cannot be covered by sea, air or ground systems. Currently, more than 1300 satellites are in orbit, of which more than 700 are communication satellites. Frequency and orbit are valuable strategic resources. The China Satellite Corporation plans to launch the ZHONG-XING 16 and ZHONG-XING 18 Ka broadband satellites by 2020. China Asia-Pacific Satellite Corporation plans to launch three Ka-band broadband satellites with a total capacity of 15–30 Gbit/s by 2020. Many similar plans exist. System isolation, information separation, system formation and service lag are problems that need to be solved.

The broadband satellite market is vast. By 2021, 4.6 million new users of broadband satellite communications will be added globally for a total of 8.1 million users. Mobile operator base station relays and emergency backups, airborne vehicle-borne and ship-borne communications, maritime communications, remote area communications, enterprise networking, regional television live broadcasting, high-definition video acquisition and distribution, personal broadband access services are very active. Faced with the rapid development of the global satellite Internet, establishing atmospheric laser communication links between satellites and earth and realizing high-speed links between satellites and earth and between satellites and between satellites is possible. Realizing the integration of multi-satellites, the multisatellite network, sky and earth, the star network, inside and outside, and communication and telemetry is the trend and goal.

5.2 Data Communication

As of December 2018, the number of internet users in China reached 829 million (21.3% in the world) (620 thousand in 1997), which is an annual increase of 56 million 530 thousand, a penetration rate of 59.6% (0.03% in 1997), which is an increase of 3.8% from the end of 2017. As of August 2018, China's mobile phone users reached 788 million, accounting for 98.25% of the total number. As of June 2018, 71% of China's online shopping and online payment users and 71.9% of mobile phone netizens used mobile payments (cited from CNNIC). The total number of fixed internet broadband access users in three basic telecommunication enterprises in China reached 407 million (37.8% of the world), with a net increase of 58.84 million for the entire year. The number of FTTH/O users is 368 million, which accounts for 90.4% of the total number of fixed internet broadband access users, which is 6.1 percentage points higher than that at the end of the last year. Broadband users continue to migrate to higher speeds. The total number of fixed Internet broadband access users with access rates of 100 Mbit/s or above has reached 286 million, which accounts for 70.3% of the total number of fixed broadband users—31.4 percentage points higher than that at the end of the last year. IPTV subscribers increased by 27% over the end of last year, for a net increase of 33.16 million for the entire year. The net increase of IPTV subscribers accounted for 44.6% of the net increase of optical fiber access subscribers.

5.2.1 Technical Progress

Data communication has experienced three stages: packet switching, internet and mobile internet. Currently, the challenges of the Internet are caused by security assurance, quality of service assurance, mobility, real-time and manageability. In recent years, to solve the problems of rigid network structure, complex closure, lack of flexibility and expansibility, and low deployment efficiency, people have proposed SDN/NFV technology, which plans to change from complex closed system to open and new open source network architecture, from administrative system to internet thinking, from passive adaptation to active, fast and flexible response. Specifically, promoting the separation and decoupling of software functions from hardware platforms; cloud and open source software functions; shift from central office (CO) centralized networking to cloud DC centralized networking; unified architecture, open network capabilities, flexible application invocation, centralized management and control; shift from operator-defined business to user-defined business; and shift from traditional vendor trading relationship to user-defined business is necessary. To construct the ecosystem of industrial chain. New breakthroughs have been made in its research, and the emphasis on standards and open source has become an important development model. Currently, more than a dozen international standardization organizations have performed SDN-related standardization

work, such as Opendaylight, ONOS and ONAP. From 2004 to 2016, Google Data
Center began to customize open source hardware. In the first half of 2018, the Linux
Foundation launched the establishment of Disaggregated Network Operating
System (DANOS) in conjunction with relevant manufacturers to create an open and
flexible network control plane. AT&T announced in October 2018 that it will submit
technical specifications for white-box base stations and gateway routers to Open
Compute Project (OCP). The StarlingX edge computing project of OpenStack is
also officially released. DPKs of Intel, SONIC of Microsoft, P4 of Barefoot and
other projects that focus on data plane performance have also been taken seriously.
ETSI dominates the study of NFV. In 2014, the United States launched the Future
Internet Architecture Plan, which focuses on cloud computing, mobility, network
infrastructure (name-based routing) and other directions. Next-generation Internet
is deployed on a small scale in the United States and China. Software-centric reconfiguration, on-demand service network, data center reconfiguration, open source
network operating system, open source network hardware design, service customization network, full-dimensional definable network architecture and baseline technology, data-driven network, multi-domain network coexistence, sharing and
interconnection are currently known as multiple candidate directions. Ultimately,
these directions will help operators to realize intelligent operation decision, customized business, precise maintenance and intelligent service.

In the future, network research will be vigorously promoted, and achievements
will be made in theoretical method innovation, international standards and application demonstration in China. China has proposed the Public Packet Data Network
(PTDN), formulated ITU-T series standards of ITU, and built a small-scale experimental network. Our country takes the lead in completing the Internet Deep Packet
Resolution series of international standards, which is the main technology in the
field of Internet Business Perception, Network Perception and Dynamic Data
Mining. This technology covers all aspects of Deep Packet Resolution technology
and products. The service-oriented future network architecture proposed by our
country effectively combines the SDN technology concept with a content routing
mechanism and constructs a small-scale experimental verification platform. The
concept of "network reconfiguration" proposed by our country explores new network architecture, addressing, routing, security and virtualization and forms a number of research results. Chinese enterprises actively participate in the activities of
international open source organizations in the field of data communication and
gradually have an important role.

5.2.2   Trend Prediction

IP data network will develop toward "Ts-abc". The internet with IPv4 architecture
as the core has some problems, such as insufficient address number, lack of security
and trustworthiness mechanism, difficult quality of service, and weak network

management ability. To achieve a controllable and secure network, the new network architecture has become an important research direction.

Network architecture is undergoing four major transformations, namely, from a complex closed system to an open and open source new SDN/NFV cloud network integration architecture, from administrative system and traditional networking thinking to Internet thinking, from passive adaptation to changes in users and customers to active, rapid and flexible response, and from traditional business relationship to construction industry as the source of elements of network architecture. Chain new ecosystem transformation. SDN/NFV, network cloud and edge computing have become the key innovative directions, promoting the network to be open, intelligent, ubiquitous and integrated. Global network architecture is turning from vertical architecture to horizontal opening, which is primarily embodied in the separation of control and forwarding, decoupling of network hardware and software, standardization of network hardware, virtualization of network virtualization and network functions, cloud and IT of network. The representative technologies are SDN and NFV [10]. SDN standardization is rapidly progressing, and NFV requirements, architecture and application scenarios have been analyzed. Cloud architecture, hardware standardization and "white box" brought by SDN/NFV, and open and open source software code have a substantial impact on the direction of data network technology and industrial development mode.

High-speed and large capacity are the long-term development direction of data communication equipment. After 400 G devices are deployed on the network, 1P devices will be gradually introduced (12.8 T bit/s single chip routing switching system has been mass produced). China will strengthen the development of large-capacity cluster routers, gradually expand the high-end router market, and enhance international voice and technology dominance by participating in standardization and open source communities.

5.3   Optical Fiber Communication

In 2018, the length of new optical cable lines was 5.78 million kilometers, and the total length of national optical cable lines reached 43.58 million kilometers. By the end of 2018, the number of internet broadband access ports had reached 886 million, a net increase of 110 million compared with the end of last year. The number of FTTH/0 ports increased by 125 million, reaching 780 million, and the proportion of access ports increased from 84.4% at the end of last year to 88%. Compared with the end of last year, the number of xDSL ports decreased by 5.78–16.46 million, and the proportion of internet access ports decreased from 2.9% at the end of last year to 1.9%. The total number of fixed-line telephone users in China is 182 million (19.3% of the world), which is 11.51 million fewer than that at the end of last year. The penetration rate is 13.1/100 (quoted from Ministry of Industry and Information).

5.3.1 Technical Progress

Optical fiber communication system has experienced time division multiplexing TDM (PDH, SDH), wave division multiplexing (WDM) and statistical multiplexing (IP transmission, MSTP, PTN, IPRAN) application scenarios.

The research progress of optical communication at home and abroad is remarkable. The transmission bandwidth is basically expanded by 1000 times of capacity in 10 years. Ground communication networks are based on optical transport networks. Mobile communication networks are also based on optical transport networks in addition to access segments. Optical network technology has become an important symbol for measuring a country's comprehensive strength and international competitiveness. Photoelectric devices are the core technology of optical network systems. Currently, the realization of single wavelength 1 T and single fiber 100 T is a relatively controversial system experiment. The international commercial DSP has developed to more than 1.25 GHz 8 core, 320 GMACS, and 160 G Flops level. As early as July 2015, 92GSa/s and high analog bandwidth ADC and DAC (28 nm process, more than 200 million gates) have been published internationally. Soon, 128 GSa/s ADC and DAC will be reported. The level of 14 nm process for FPGA has reached 550 million gates. Currently, 400 G bit/s optical fiber transmission technology has been mature for commercial use, and the industry is formulating international standards for rates greater than 400G bit/s. After 10G PON, the single wave 25G PON is gradually applied, and the realization of 2 *25G or 4 *25G PON by wavelength superposition is also placed on the application agenda. The impact of SDN/NFV and optical network cloud integration on optical communication is gradually emerging, and the level of intelligent optical network is increasing, which shows a rapid development trend. Silicon photon integration and III-V material integration technology have been accelerated, passive optical device integration has been greatly developed, and 25G active integrated optical modulator/detector has made breakthroughs.

In recent years, optical communication technology has rapidly changed from follow-up mode to innovation-driven, which constantly impacted the international leading level. By the end of 2016, China has realized 200 Tbit/s ultra-large capacity WDM and FDM optical fiber transmission, breaking through the visible space channel interference suppression technology, and the visible real-time communication rate experiment has been increased to several Gbps. A silicon-based integrated 100G bit/s coherent modulation and reception chip is developed. The working speed of the silicon light modulator is increased to 80G baud. The photoelectric integrated sample of 32G bit/s silicon light modulator and CMOS driver is developed. In the research of quantum key distribution, quantum teleportation with pre-entanglement distribution between independent quantum sources is realized.

From the point of view of scale commercialization, the rate chips for over 100G/400G are the key, including InP series (high-speed direct modulation DFB and EML chips, PD, APD, high-speed modulator, multichannel adjustable laser chips, etc.), SiP series (coherent optical transceiver chips, high-speed modulator, optical switch chips, TIA, LD Driver, CDR chips, LiNbO 3 series (high-speed modulator chips), GaAs series (high-speed VCSEL, pump laser chip), and Si/SiO_2 series (PLC, AWG, MEMS chip).

5.3.2 Trend Prediction

Fiber optic communication will develop towards IPs-abc. In recent years, we focus on 400G bit/s and 1 T bit/s high-speed optical transmission, 5G forward/return and high-precision synchronization, 100G-PON and WDM-PON optical access, optical network SDN/NFV. We further grasp the new optical fiber preform rod and achieve advanced level in the preparation of new optical fibers, such as large effective area, ultra-low loss. Consider the silicon photon as an example, by gradually grasping the high-speed modulator, and silicon-based switching array. The key technologies, such as silicon optical chips, gradually solve the problems of low localization rate and lack of core technology of high-end chips. Based on the existing 10 M, 100 M, GE, 10GE, 25GE, 40GE and 100GE rates, the transmission rate of Ethernet will evolve to 400GE and 1TE and higher rates, breaking through the optical interconnection of single channel 400 Gbps/1Tbps data center, integrating density not less than 500 $Gbit/s/cm^2$ and 1 $Tbit/s/cm^2$, and adding flexible bearing characteristics, such as interface channelization, and network fragmentation, such as high speed, large bandwidth, low delay, high reliability and high security are main trend.

With the goal of "P-bit optical fiber transmission, photoelectric integration and all-optical interconnection", China's optical communications will focus on core technology research, enhance international voice, and break the monopoly of developed countries on the core technology of optical communications. New technologies and multiplexing technologies, which greatly enhance transmission capacity, have become the focus of research. Research on future-oriented new space-division multiplexing technology, such as silicon-based optoelectronic integration, new code modulation, ultra-dense wavelength division multiplexing, multicore (MC) multiplexing, and small-mode (FM) multiplexing. Exploring new fiber materials and new process technologies, breaking through the core technology of photonic integration and optoelectronic integration, and completing the research and production of submarine cable systems, ultra-low loss optical fibers and super large core fibers, laying the foundation for building China's optical communication core technology.

5.4 Mobile Internet

In 2018, the mobile internet access traffic reached 71.1 billion GB, which is an increase of 189% from the previous year, and an increase of 26.9 percentage points from the previous year. The annual average monthly mobile internet access traffic (DOU) was 4.42 GB/month, 2.6 times that of the previous year, and 6.25 GB/month in December 2018. The mobile internet traffic reached 70.2 billion GB, which is an increase of 198.7% from the previous year and accounts for 98.7% of the total traffic (cited from Ministry of Industry and Information Technology).

5.4.1   Technical Progress

With the people as the center, the shift of communication mode from the minority to the public is an important feature of mobile internet. In recent years, China's mobile internet has witnessed blowout growth, and the 4G network has greatly improved. The average monthly usage time of its user traffic is second only to Wi-Fi, with more than four million APPs. Due to the continuous innovation of mobile Internet technology and business model, opportunities for change in enterprises, industries and products exist. Many business models in China are in the lead in the world, including mobile payment, WECHAT and online shopping. A number of new enterprises with core competitiveness may emerge. At the social level, this has changed the traditional mode of interpersonal communication; at the economic level, e-commerce and mobile payment have become common, human embedded micro-phones will enter commercial applications, business models can be "compared" to the industry began to change, most people will have a digital avatar on the Internet, "shared economy" is constantly rising. At the level of social management, this will change the way of social governance. The mode of social management is changing from one-way management to two-way interaction, from off-line to on-line and off-line two-way integration, from simple government management to a governance mode that pays more attention to social coordination.

5.4.2   Trend Prediction

Mobile internet and its combination with the real economy are the new trends of China's current economic development. With the continuous popularization and deepening of mobile internet technology, the trend of its application is to construct new application systems and develop new modes of innovation, open and open sources, and cross-integrations with various industries. From the perspective of the development of the mobile internet industry, first, the manufacturing power, industrial internet and intelligent manufacturing are redefining the global manufacturing industry; the shape, boundaries and modes of enterprises are changing; and products,and services are constantly integrated. Second, mobile APP and various technology industries cross integrate, extend, and penetrate into all aspects of social life, and new products and new models are developed. Third, the mobile Internet connects 1.4 billion people via mobile phones and has become a super-organized, living system that exists only through connectivity. More than one billion instances of information sharing occur through this connection every day. Through the process of upgrading, this connection will become increasingly intelligent. Through the combination of artificial intelligence, subversive concepts will constantly emerge. In the next few years, the most valuable and innovative concepts will appear here.

In future years, from the perspective of the manufacturing industry, the software platform will extend to services and pan-terminals, the proportion of service-oriented manufacturing will expand, and the focus of competition will gradually shift from pure market share expansion to high-quality ecosystem services. New products of mobile intelligent terminals will accelerate remodeling, intelligent VR/AR, three-dimensional holographic virtual technology, APP, and wearable terminals. New technologies such as end-to-end, eye mask displays and foldable displays will gradually enter public life. The boundaries of pan-intelligent terminals will continue to extend and innovate, which may form a new momentum of economic growth after smartphones.

In the next few years, from the perspective of application and business, the global economic model will accelerate to the digital economy as represented by the mobile internet, cloud computing, big data and artificial intelligence. According to the statistical results from the first half of 2018, China's digital economy with the Internet as the carrier reached 31 trillion yuan, accounting for 1/3 of its GDP. China's e-commerce transactions account for 42% of the world's total, and mobile payment transactions in China are 11 times as large as those in the United States. The scale of China's digital economy is expected to exceed 32 trillion yuan by 2020 (quoted from the White Paper on China's Digital Economic Development and Employment 2018). Compared with the traditional internet, the contribution of the mobile Internet is more prominent. The general practice of the traditional Internet is to set up a platform, provide services, rent businesses and charge moderately. The platform is the center; users access it from outside. The typical characteristics of the mobile internet are platform decentralization; user coverage; communication civilianization; service personalization; communication fragmentation; timely information release; free, flat, simplified and intelligent; constantly breaking traditional industry boundaries; forming a win-win mechanism online and offline; upstream and downstream of the industrial chain; value chain; endless upgrading; accelerating technology iteration; and resource sharing. The mobile internet has the characteristics of media, tools, interaction and community. Innovative technology originates from the combination of existing technologies, which differs from the traditional Internet. Innovative technology is public-centered, multichannel, diversified and socialized. The mode of innovative technology is from the minority to the public, and it is multidirectional and multilateral. Relying on brand-new social relations, group building and one-to-one connections can quickly form a network and a chain reaction, and the number of connections undergoes exponential growth. When I travel, I can take almost nothing with me, just a small cell phone. This round of scientific and technological revolution and industrial revolution converge at an unprecedented speed and momentum. With the application of artificial intelligence to heterogeneous and multi-scene massive human-network-material data deepening and human-computer integration, a better, new organizational structure will be developed in the next 20 years or so, which will gradually form a city and wider range of network agents, which will develop in a more, larger and faster direction and be associated with all people and most "things", followed by people and people, people and themselves. However, productivity, production relations, superstructures and human beings themselves will be redefined.

5.5 Internet of Things

More than 5000 kinds of sensors are used in the front end of the Internet of Things and are separated into 10 categories, 40 categories and more than 5000 varieties. The Internet of Things penetrates into every corner of the world through these sensors. In 2018, three basic telecommunications enterprises in China developed 671 million users of the cellular Internet of Things, a net increase of 400 million in the entire year. The continuous progress and application of Wide Area Wireless Connection technology promote the rapid development of Internet of Things technology. Low power consumption, wide coverage, small data and ubiquitous connection are the remarkable characteristics of Internet of Things technology. The IoT uses several main technical paths. The first path is based on the optimization of existing cellular network technology, mainly EC-GSM, TD-LTE Cat.M and FDD-LTE Cat.M technologies. The second path is oriented to the wide coverage and low coverage of the Internet of Things. The third path is the 5G technology standard for delay scenarios, which is low power wide area network technology (LPWAN). One example is authorized wide area network technology, mainly referring to NB-IoT and the LTE evolution technology eMTC, and another example is non-authorized wide area network technology, referring to LoRa, PRMA and Sifox. Different LPWAN technologies have different technical requirements, deployment methods and service modes. New identity resolution systems, Internet of Things semantics technology and interoperability are also controversial topics of research.

5.5.1 Technical Progress

New identity resolution systems, IoT semantic technology and interoperability have become research hotspots. The main international identification systems of the IoT are Electronic Product Code (EPC), Object Identifier (OID), and Ecode. To solve the problem of resource interoperability caused by network heterogeneity and cross-systems in the Internet of Things, semantic technology is introduced. Many semantic models have been proposed to create connections between data. OnM2M abstracts different devices into resources; establishes a unified interface between resources; realizes the creation, acquisition and search of resources based on an open protocol framework; and discovers other related data via one source of data. The key to the semantic interoperability of the Internet of Things is to establish data models and analysis tools to provide cross-system knowledge sharing and automatic application.

China has initially proposed the architecture and model of the IoT, laid out the IoT, and initially established the public service platform of the logo management of the IoT, which is conducive to the interconnection of different logo analysis systems. Research institutions, universities and major enterprises in China have increased their investment in standardization, which has become one of the major contributors to the ISO and ITU Internet of Things Working Group. This working group has formulated the "Overview of the Internet of Things" and the Guidelines for the integrated standardization system of the Internet of Things. This working group has sorted out hundreds of standard items. In terms of the internet of things reference architecture, intelligent manufacturing, IoT Semantics, evaluation of electronic health indicators and big data, the main relevant international standards for the IoT of China have been issued. Among these standards, China actively promotes the development of global NB-IoT standards and industries. In fields related to the Internet of Things of OneM2M, 3GPP, ITU, IEEE and other major standardization organizations, China has won more than 30 leading positions in standards organizations related to the Internet of Things, presided over relevant standardization work, and has been able to provide services in the wireless WAN of the Internet of Things, Web-based Internet of Things service capabilities, and vehicles. Networking, wearable equipment and other aspects have formed a common leading standard-setting trend with developed countries.

5.5.2   Trend Prediction

3-abc is a dominant trend, where "3" refers to the three hundred billion-level interconnection of people-network-things and the integration of 5G/4G-LTE IoT technology and technologies of various industries, "a" refers to the artificial intelligence (AI) of wireless mobile networks, "b" refers to the big data of wireless mobile networks, and "c" refers to the cloud of wireless mobile networks; all services are user-centered. 3-abc includes in-depth learning, intelligent analysis and processing, intelligent perception, intelligent protection of security incidents and situational awareness. Using a soft definition of virtualization and intellectualization, all things can be interconnected, connected, extended and penetrated into the vast corners of the world, changing things from "dead" to "alive", all under human monitoring and scheduling (through online information). 3-abc can be used to create new needs, new models, new concepts and new possibilities. Considering global industrial systems as an example, their integration with the intelligent cloud, advanced computing, analysis, perception technology and Internet system causes the emergence of an industrial Internet system that links people, data, products and machines; forms intelligent equipment, intelligent systems and intelligent decision-making; upgrades the traditional manufacturing industry to intelligent manufacturing; and forms a new production mode, organizational structure and business model type, industrial form and economic growth point. In line with this trend, hundreds of billions of sensors will be connected to the Internet over the next 10 years. McKinsey estimates that the impact of the Internet of Things on the global economy will reach $2.7 trillion to $620 billion by 2025. Accenture predicts that with its uncertain economic growth prospects, the Internet of Things is expected to contribute to the global economy by 2030 with a new output value of $1.42 trillion, which will lead to substantial changes to infrastructure construction, energy, health care, manufacturing, materials, and supply chain management.

The application of the IOT is rapidly advancing throughout the world and is moving from the start-up stage of fragmentation and isolation to the new stage of intensive "unified top-level, focused focus, cross-border integration and integrated innovation". The Industrial Internet has become a new research hotspot. The application of intelligent transportation, intelligent medical treatment, intelligent education, smart homes, smart enterprises and vehicle networking has gradually matured and undergone accelerated penetration, and the global smart city has entered the stage of scale development.

The Narrow Band Internet of Things (NB-IoT) has become an extension of the wireless Internet of Things, relying on the existing mobile network via big connections, wide coverage and low power scenarios. NB-IoT is expected to end the phenomenon of "isolation and fragmentation" of past Internet of Things applications. In addition, eMTC, which has a higher speed than NB-IoT, has attracted the attention of operators in North America and Japan and is also an important development direction of low power wide-area interconnection technology. Many provinces and municipalities in China have begun to deploy NB-IoT for sharing bicycles, smart water meters, smart gas meters, smart parking lots and so on. In 2018, it NB-IoT is expected to add 300,000 base stations for network coverage. By 2020, the scale of NB-IoT base stations in China is expected to reach 1.5 million and achieve national coverage. Tunnel/oil depot fire detection, bridge safety protection, optical fiber sensing (passive, battery-free), safety fence are also excellent options for IoT solutions [11].

There are at least three obstacles to the development of the Internet of Things in the near future: first, there are many kinds of technical standard paths, none of which has absolute advantages at present, and the follow-up will be clear with the application of technology and industrial development. Second, the battery endurance of Internet of Things terminals is insufficient, and further technological breakthroughs are needed. Third, the issue of information security will become the focus of large-scale applications of the Internet of Things.

5.6 Edge Computing

The world is undergoing a wave of digital transformation that is affecting the mode of industrial manufacturing; people's daily communication; the travel, education, medical industries; and other traditional services. Edge computing is defined as a distributed open system that can lead to the construction of a convergence network near the edge of objects or data sources, computing, storage and core capabilities of applications. Edge computing provides near edge intelligent processing capabilities to meet the key requirements of agile connections, real-time businesses, data optimization, application intelligence, security and privacy protection. Gartner predicts that by 2021, 40% of large enterprises will incorporate edge computing into their projects due to time delays and bandwidth requirements. According to IDC statistics, the total amount of data accessed worldwide will exceed 40 TB by 2020, of which 45% of the data generated by the Internet of Things will be processed at the edge of the network.

5.6.1   Technical Progress

The application goal of the development of edge computing is to provide computing, storage and connection as basic capabilities to all industries and realize digitalization, networking and intellectualization of the industry. The main technical concepts are hardware generalization and software-hardware decoupling, open computing and storage capacity, task scheduling by software, and real-time processing capability on demand.

One way to realize edge computing is to use sink ICT infrastructure to provide richer computing power at the edge side. The main object-orientation are consumers and industry users, with the operation of enterprises as the leading factor. A typical implementation is the deployment of the edge cloud. A traditional centralized data center strategically provides advances to small data centers, providing users in different industries with on-demand computing. For example, at the end of 2017, China Unicom, Tencent and Intel jointly established a test bed for edge data centers and plan to build 6000 edge data centers over the next few years using this technology to provide related services. Another technical path of edge computing is to provide an open interface for third-party service applications by upgrading field-side devices, of which the most typical application is the edge gateway. For example, Amazon released "AWS Greengrass" edge computing software, which seamlessly extends the functions of Amazon Cloud Services to industrial devices; Aerospace Cloud launched an industrial Internet of Things gateway product that can connect to its INDICS industrial Internet platform, providing the ability to collect, convert, process and transmit data from industrial devices from different manufacturers brands, as well as inside and outside the factory. One way to realize edge computing is network and communication protocol conversion functions. Established in 2017, Edgecross Alliance of Japan, which consists of Mitsubishi Electric, Yanhua, Omron, Japan Electric, Japan IBM and Japan Oracle, launched an open platform and application in 2018 to provide basic edge computing software services.

The key technologies of edge computing include virtualization technology, a software definition network, a time sensitive network (TSN), heterogeneous computing (HC), a sequential database (TSDB), etc. According to several key aspects of edge computing, edge computing platforms are usually designed for specific types of computing scenarios, and their load types are relatively fixed. Therefore, there is still a pioneering need regarding the design of the framework of edge computing platforms for specific computing schemes. The use of dedicated acceleration units reduces the execution time of one or more types of loads and significantly improves the performance and power consumption ratio. Unlike the real-time operating systems Contiki and FreeRTOS on traditional Internet of Things devices, edge computing operating systems need to manage heterogeneous computing resources downward, process large amounts of heterogeneous data and multiple application loads upwards, deploy complex computing tasks on edge computing nodes, schedule and migrate tasks to ensure their reliability and maximize resource utilization. Current technical research focuses on the management framework of data, computing tasks and computing resources.

5.6.2   Trend Prediction

Since the concept of edge computing was proposed a few years ago, it has undergone explosive growth. According to its trend of rapid development, edge computing will produce a greater spillover effect in the future and will become an adhesive that connects all walks of life and a catalyst to promote the upgrading and transformation of the whole industry. In the development trend of technology, mass terminal connections will have a far-reaching impact on artificial intelligence, machine learning and other technologies. Mass terminal connections will promote the development of micro-kernel technology and promote the embedding of algorithms and models into the firmware of mass devices, so that front-end intelligence will have more development prospects and edge intelligence will be an important technical support. In the application development trend, edge computing has formed an industrial consensus. The development of the Internet of Things, Mobile Internet and Industrial Internet will make the data processing platform develop in the direction of "decentralization". Border cloud collaboration will become the system construction mode of future industry digital transformation [12].

At the same time, industrial alliances and open source communities play an increasingly important role in promoting the development of marginal computing and building industrial ecology. In February 2019, the Industrial Internet Consortium (IIC) and the OpenFog Consortium announced their merger. IIC and the OpenFrog Consortium will work together under the framework of IIC to develop and promote industry guides for fog computing and edge computing as well as industrial practices. In addition, the technology community has also provided support for the development of edge computing, in which Kubernetes will be widely used in ultra-large-scale cloud computing environments. To promote the application of Kubernetes in edge scenarios, the new Kubernetes Edge Computing Working Group will adopt container architecture and extend it to edges.

The development of edge computing is in a critical period of technological innovation. Various industrial camps are also occupying a dominant position in edge computing, and it will experience many challenges in the future. The technological challenges of edge computing include the urgent need to unify the architecture; the immature technology theory; the need for an extensive range of equipment; heterogeneous data; the lack of open standards; the intelligent situational awareness capabilities; and improvement of the unified open platform. Application challenges include: the need to enhance the openness, versatility, compatibility and security of edge computing platforms; the fragmentation of technical solutions; the difficulty of integrating cross-vendor interconnection Technical solutions; and the need to explore new business models of multiparty collaboration.

5.7   Quantum Communication

Quantum Teleportation (QT) and Quantum Key Distribution (QKD) are two main components of quantum communication. Quantum teleportation still has many technical problems that need to be solved and is a hotspot in the field of basic scientific research, but there is no clear prospect for its practical use [13]. The main focus of the research on quantum key distribution is to improve the level of system integration and practicality and explore new application scenarios. In the field of quantum communication, there has been a series of research achievements made by research institutions, such as China University of Science and Technology and Tsinghua University, which basically keep pace with the international advanced level. The layout of scientific research and practical exploration in the field of satellite-to-ground quantum communication in China is in the relatively early stage. In 2016, a quantum science experimental satellite (Mozi) was successfully launched, and research results were achieved in the field of satellite-to-ground quantum entanglement distribution, teleportation and key distribution. In the pilot application of the QKD-based quantum secure communication system, several demonstration verifiable circuits and metropolitan area networks for quantum secure communications have been built in Beijing, Hefei, Wuhu and Jinan since 2007. Since 2017, the Quantum Secret Communication "Beijing-Shanghai Trunk Line" project has been running and passed acceptance, and the "Shanghai-Hangzhou Trunk Line", "Nanjing-Suzhou Trunk Line" and "Wuhan-Hefei Trunk Line" projects have begun to be built. At the end of 2018, the first phase of the national wide-area quantum secure communication backbone network construction project was implemented, which provided an important impetus for promoting the application and industrialization development of quantum secure communication.

5.7.1   Technical Progress

Direct information transmission is based on the entanglement distribution, Bell state measurement and unitary transformation between the two sides of the communication. The transmission of quantum state information still requires the assistance of traditional communication methods. All kinds of reports on QT experiments are limited to proving the feasibility of the principle and observing experimental phenomena, and the practicality of using quantum communication distance is based on QT. The entangled light source used in QT is usually fabricated by a combination of a laser and nonlinear crystal. The generation of entangled photon pairs requires a posterior probability process based on measurement verification. The generation efficiency and application scenarios are limited. The practical prospect of a high-quality deterministic entangled light source remains unclear. In addition, entangled photon pairs are susceptible to decoherence due to environmental noise and quantum noise in the process of a distributed transmission. The quantum entanglement characteristics are difficult to maintain. Quantum relay based on quantum state storage and entanglement switching technology can overcome the decoherence problem in the quantum entanglement distribution and increase the transmission distance. However, there are advantages and disadvantages in terms of the storage time, fidelity, storage capacity and efficiency of various technical schemes for quantum state storage, such as using a gas cold atom ensemble, rare earth ion doped crystals and QED cavity atom trapping. However, there is no one technical scheme that can meet the practical requirements of all the indicators at the same time. Quantum storage and quantum relay technology need to be investigated.

In QT experimental research, the Dutch validated that the coherence of long-range entangled electrons exceeded the Bell inequality limit based on the diamond color center system in 2015, which fully proved the nonlocality of quantum entangled states. In the same year, the United States and Japan jointly realized the longest distance quantum teleportation experiment of 102 km in optical fibers. In 2018, the entanglement effect between macroscopic materials was observed by using a 1550 nm entangled photon pair interaction between nanostructured optical resonator chips at the University of Vienna, Austria. In 2017, China University of Science and Technology reported the Mozi satellite-to-ground QT transmission experiment. Uplink was used between the LEO satellite and ground station. The longest reported transmission distance of the quantum entanglement distribution and teleportation was 1400 km.

The QKD protocol can be divided into the discrete variable (DV) protocol for single photon modulation and the continuous variable (CV) protocol for regular component modulation of the light field according to the different coding modes of quantum state information. DV-QKD includes the BB84 protocol, differential phase shift (DPS) protocol and coherent one-way (COW) protocol, among which the BB84 protocol is the most mature protocol with more complete security certification and a higher commercial level of the system equipment. The BB84-QKD device integrated with decoy state modulation can be further subdivided into polarization modulation, phase modulation and time-phase modulation according to the different modes of the single photon quantum state modulation and demodulation. The post-processing process of the BB84 protocol includes five steps: sifting, error estimation, error correction, confirmation and privacy amplification. Error estimation and encryption enhancement are the core steps to ensure the security of QKD, and the efficiency of the error correction and verification algorithm is one of the bottlenecks that limit the secure coding rate of QKD. In QKD metropolitan area networks, wavelength division multiplexing or optical path switching (time division multiplexing) between the quantum state optical channel and synchronous optical channel is usually realized by combining dividers or optical switches. Due to the immaturity of quantum relay technology, the long-distance transmission of the QKD optical fiber system can only rely on the trusted relay technology of the key landing and hop-by-hop relay. The key storage management and relay forwarding of trusted relay nodes that are needed to satisfy the relevant requirements of the cryptographic industry standards and management standards and sites usually need to meet the relevant requirements of the information security level protection or have corresponding security protection conditions.

According to the transmission and measurement of quantum states, QKD achieves secure key sharing that cannot be eavesdropped between the sender and the receiver and then combines with traditional secure communication technology to realize the encryption and decryption of classical information and a secure transmission. In 2018, the University of Geneva, Switzerland reported that an ultra-low dark recording rate superconducting nanowire single photon detector was used to realize long-distance QKD transmission in a 421 km ultra-low loss optical fiber system, but the key coding rate was only 0.25 bit/s. In the same year, Toshiba Cambridge Research Institute reported an experiment using a high key coding rate system based on T12-QKD protocol and LDPC error correction coding. The system ran continuously for 4 days on a 10 km optical fiber line, and the average key coding rate reached 13.72 Mbit/s. Secure communication based on quantum key distribution is called quantum secure communication. The protocol technology and equipment system used for quantum secure communication are relatively mature and mainly serve the information security field. At present, quantum secure communication has entered the preliminary industrial application stage.

5.7.2 Trend Prediction

Quantum communication and the quantum Internet based on QT will be the frontier research hotspots in the field of quantum information technology in the future. The NQI Act of the United States lists safe communication based on QT as one of the four major applications, as well as large-scale interconnection and information communication of quantum computers through the quantum Internet. The flagship of the European Quantum Declaration plans to establish the Quantum Internet Alliance (QIA) in the first batch of projects to support research institutions, such as

Delft University of Technology in the Netherlands; develop quantum communication terminals and repeaters; and establish a quantum communication experimental network supporting quantum bit transmission and networking among four cities in the Netherlands by 2020.

In the future, the development direction of the evolution of QKD technology primarily includes three aspects: enhancing system performance, enhancing realistic security and improving the level of practicality. For example, the new protocol for high-dimensional quantum state information coding (HD-QKD) and phase random Double-field (TF-QKD) is adopted; the new MDI-QKD protocol is adopted to eliminate detector-related security vulnerabilities and improve the real security level; and common-fiber transmission and fusion Networking Research of the QKD system and the classical optical communication system is performed. Research on common fiber transmission and fusion networking of the QKD system and the classical optical communication system, the QKD device chip based on photonic integration technology and the handheld intelligent terminal will further improve its practical level.

Chapter 6
Annual Hot Words

6.1 Hot Word 1: 5G

Basic Definition: 5G is the abbreviation for fifth generation mobile communication, which is next generation wireless communication technology with an ultra-high bandwidth, low delay and vast connections. This technology is a new milestone in the development of mobile broadband technology after 4G. This kind of communication technology will be closely combined with artificial intelligence, big data, and the Internet of Things. 5G will initiate leapfrog innovation in many industries, profoundly change people's production and lifestyles, and begin a new digital era of the interconnection of all things [14].

Application Level: In February 2018, the first 5G 3GPP standard commercial chips and terminals were released at the Mobile World Congress and 5G smartphones were introduced to the market. Operators in the United States, Japan and South Korea have also opened 5G pilot networks that are in the pre-commercial stage, which provide voice, mobile broadband, remote control and other related 5G services. These operators each hope to become the first country in the world to provide 5G commercial networks. 5G technology development in China is at the forefront of the world and is competitive. The first batch of 18 5G pilot cities in China are gradually focusing on the construction and development of 5G commercial networks. The 5G era has arrived. For example, the Hankou District in Shanghai plans to build a 5G network in the administrative region in China, which will achieve a Gigabit fixed broadband network and 5G network coverage.

© China Science Publishing & Media Ltd (Science Press) 2020   43
Center for Electronics and Information Studies, Chinese Academy of Engineering,
Network and Communication, https://doi.org/10.1007/978-981-15-4596-2_6

6.2   Hot Word 2: 5G New Radio (5G NR)

Basic Definition: The new radio refers to the connection protocol between a mobile terminal and a base station. A unified 5G adaptive new radio port must have the ability to respond to complex and diverse business application requirements and flexibly adapt to various businesses. 5G-NR includes a new waveform (DFT-s-OFDM, CP-OFDM), modulation mode (1024QAM), frame structure, multiple access technology (SCMA, NOMA), coding technology (Polar Code, LDPC), multi-antenna transmission and a series of new wireless port technologies [15].

Application Level: In June 2018, the International Organization for Standardization of Communications (ISO) 3GPP froze the 5G-NR SA protocol. The freezing of the NSA and SA standard indicates completion of the first stage of 5G standardization and the official release of global complete 5G international standards. This progress is the result of the cooperation and compromise of major global manufacturers and has been extensively recognized. Many communication equipment manufacturers have released pre-commercial products based on 5G-NR and demonstrated 5G-NR multivendor interoperability that is consistent with the 3GPP standard.

6.3   Hot Word 3: 5G Carrier

Basic Definition: A carrier network is a private network built by operators to carry various voice and data services (such as soft switching, video, and key customer VPN) usually using optical fiber as the transmission medium. The 5G network is moving forward. As an end-to-end network that connects wireless and core networks, a carrier network is divided into three parts: access layer, convergence layer and core layer. The access layer corresponds to the 5G forward and intermediate transmission network, and the convergence layer/core layer corresponds to the return network.

Application Level: In China, the IMT-2020 (5G) Group has taken the lead in formulating two pilot plans, including the technology R&D test (2015–2018) and product R&D test (2018–2020). Currently, no unified scheme for the choice of 5G carrier technology exists in the industry, and each of three operators has unique advantages. The China Mobile leading SPN scheme has built a corresponding SPN experimental network and started to promote it. China Telecom, which has the leading M-OTN scheme, has achieved two related standards. China Unicom, which has the leading metropolitan WDM scheme, has led the adoption of the ITU-T G.698.4 standard. Because a wired carrier network needs to be deployed earlier than a wireless network, the 5G carrier network technology of many enterprises is maturing. Therefore, in 2019, the 5G carrier network will be the first network to complete deployment while waiting for the arrival of 5G commercial.

6.4   Hot Word 4: AI+

Basic Definition: On March 5, 2019, Premier Li Keqiang presented the concept of "AI+" in the government work report presented at the second session of the 13th National People Congress. The significance of this report is the use of artificial intelligence technology to empower all walks of life, that is, the use of cloud computing, big data, the Internet and other new generation information technology. Thus, AI technology can be deeply integrated with traditional industries to create emerging industries. The new model and new format represents a new social form, that is, to highlight the role of AI in society, apply its innovative achievements to industrial production and all aspects of social life, comprehensively enhance innovation ability and work efficiency, and form a broad new form of development based on the Internet, Internet of Things, Industrial Internet and other infrastructure and enabling tools.

 Application level: Currently, China is in the transition stage of "Internet+" to "AI +". In the era of "Internet+", technology applications are primarily aimed at consumers, and more applications will be geared to end users. Information technology application is not limited to people. People and things in addition to things and things can be interconnected using more intelligent machines, more intelligent networks, and more intelligent interactions to create a more intelligent economic development model and social ecosystem. In March 2018, unmanned buses were officially placed in operation, and 4 months later, L4-class mass-produced self-driving buses were officially launched in China. In April 2018, the domestic "unmanned bank" officially opened in the Shanghai Branch of the China Construction Bank. In November 2018, at the Fifth World Internet Congress, a full intelligent virtual host "AI synthetic anchor" officially appeared. Numerous no-one hotel, no-one restaurants and factories were constructed. AI + transportation, AI + medical treatment, AI + finance, AI + education, AI + manufacturing and other applications are accelerating the development. A new era of AI + is emerging.

6.5   Hot Word 5: Industrial Internet

Basic Definition: The Industrial Internet is based on the interconnection of machines, raw materials, control systems, information systems, products, data and people. The Industrial Internet integrates the global industrial system with a new generation of information technology, such as supercomputing, comprehensive perception and the Internet. The Industrial Internet is an important carrier of digitalization, networking and intellectualization of the manufacturing industry. The Industrial Internet primarily includes three systems—network, platform and security—and realizes the deep integration and integrated application of information and communication technology in all industrial elements, fields, industrial chains and value chains.

Application Level: The Industrial Internet is the common choice for major industrial powers to achieve intelligent manufacturing and seize the competitive heights of the international manufacturing industry. In February 2018, China set up a special working group on the Industrial Internet with a leadership group to build a strong manufacturing country. In June 2018, the Ministry of Industry and Information Technology issued "the Industrial Internet Development Action Plan (2018–2020)" and "the Industrial Internet Work Plan 2018". Currently, the application of the industrial Internet platform concentrates on three scenarios: equipment management service, production process control and enterprise operation management. The optimization of resource allocation and product R&D design have been initially applied but the total needs have to be nurtured. Driven by the law of application value, the application of the industrial Internet platform presents three levels of development: hot spot depth data analysis, cloud resource docking and data mechanism precipitation exploration. Some industrial enterprises have performed the practice of the industrial internet, and enterprises such as the internet, telecommunications and software are actively building the industrial Internet ecosphere. Generally, the development and application of the industrial Internet remains in the initial stage.

6.6   Hot Word 6: Artificial Intelligence Internet of Things (AIoT)

Basic definition: AIoT = AI + IoT. AIoT is the technological integration of AI and the Internet of Things in a practical application. AIoT is not a new technology but a new application form of the IoT. Through the IoT, a massive amount of data is generated, collected, processed, and stored in the cloud and edge. Multisource heterogeneous large data technology is employed for analysis and artificial intelligence decision-making to achieve all things data and all things intellectualized. AIoT pursues an intelligent ecological system based on the Internet of Things. AIoT differs from traditional IoT applications. If the objective of the traditional IoT is to connect all ordinary objects, which can perform independent functions and connect all things with the network, then AIoT is endowed with more intelligent features to achieve the true sense of all things in the wisdom of association.

Application Level: The development path of AIoT will undergo three key stages: single machine intelligence, interconnected intelligence and active intelligence. Currently, the AIoT industry is only in the stage of single aircraft intelligence. Over the next 20 years, AIoT will become one of the most important technologies in the world and the important foundation of new industries, such as industrial robots, unmanned aerial vehicles, intelligent companionship, intelligent buildings and intelligent cities. Intelligent IoT devices will be ubiquitous, everything in the world will be interconnected, and human society will enter the era of the artificial intelligence of all things.

6.7   Hot Word 7: Golden Spectrum

Basic Definition: Usually refers to the medium- and low-frequency spectrum resources that require spectrum recultivation in the 5G era. These bands are generally in the golden frequency range. For example, the Band8 band defined in E-UTRA has an 880–915 MHz upstream band and a 925–960 MHz downstream band. In the frequency band below 6 GHz, the C-Band (3400–3800 MHz) has become the most extensively employed mobile communication frequency band in the world. Due to the excellent wireless transmission characteristics in the low- and middle-frequency band, it is a valuable resource to which major operators strive.

Application Level: The United States, Japan, South Korea and other countries are making efforts to promote the 28 GHz band for hot spot coverage and last kilometer access. China and the European Union focus on promoting the use of frequency bands below 6 GHz for wide coverage. After many balances and compromises, the scheme of the spectrum division of 5G frequency bands in China has been basically determined. China Telecom obtains 100 MHz bandwidth resources from 3.4 to 3.5 GHz; China Unicom obtains 100 MHz bandwidth resources from 3.5 to 3.6 GHz; and China Mobile obtains 260 MHz bandwidth resources from 2515–2675 MHz and 4.8–4.9 GHz. The spectrum division scheme enables a balanced competition pattern of the industry, which lays a solid foundation for 5G construction.

6.8   Hot Word 8: Millimeter Wave

Basic Definition: The telecommunication industry refers to the electromagnetic wave in the frequency domain of 30–300 GHz (wavelength 1–10 mm), also known as the millimeter wave, which is located over the wavelength range where microwaves and light waves overlap, and has the characteristics of two kinds of spectra. In 5G networks, the millimeter wave has not been precisely defined. Generally, electromagnetic waves above 24 GHz are collectively referred to as the millimeter wave.

Application Level: US operators have deployed a 5G millimeter wave fixed wireless access network in 28 GHz band and accelerated commercial deployment of millimeter wave 5G network, which conforms to the 3GPP R15 standard. Based on the demand of large-scale emerging applications, the global millimeter wave technology market will increase at a compound annual growth rate of approximately 30%. A preliminary consensus has been reached in the Asia-Pacific region that the 26 GHz millimeter-wave band will be utilized in 5G systems. Substantive progress has been made in the international coordination of the millimeter-wave band. In China, the Ministry of Industry and Information Technology approved a total of 8.25G millimeter-wave bandwidth for the 5G technology R&D test, which shows the determination of introducing high-frequency resources in China. Due to the progress of millimeter wave technology, materials and products, millimeter wave products for communication are gradually entering the commercial stage.

6.9 Hot Word 9: Terahertz

Basic Definition: Terahertz usually refers to the electromagnetic wave from 0.1 to 10 THz. In the electromagnetic wave band specified by the International Telecommunication Union (ITU), the frequency is 0.3–3 THz. This band is the last span in the entire spectrum of electromagnetic waves and is referred to as "THz Gap" in the scientific community. Terahertz is a frontier research field with an interdisciplinary nature, which is suitable neither for optical theory nor microwave theory.

Application Level: Terahertz is suitable for wireless communication systems with transmission rates that exceed 100G bit/s, such as new generation wireless local area networks and wireless personal area networks. Terahertz technology is also one of the key technologies in future 6G mobile communication. Currently, terahertz communication remains in the stage of research and development of key devices, feasibility demonstration of the total structure scheme of terahertz communication system and laboratory research and simulation demonstration. The development of high-efficiency solid-state devices, such as transmitting antennas and radiators, is needed to solve the modulation and processing technology of terahertz signals and formulate the corresponding technical standards. Terahertz communication technology can achieve higher speed information transmission and occupy spectrum resources, which has high economic value and strategic significance [16].

6.10 Hot Word 10: Quantum Key Distribution

Basic Definition: Quantum communication technology is a new type of communication method that uses the quantum entanglement of micro-particles or the quantum entanglement effect to distribute keys or transmit information. This technology is a new interdisciplinary subject that has been developed in recent years. Current quantum communication technology primarily refers to quantum key distribution (QKD). Quantum key distribution is based on the principles and characteristics of quantum mechanics to ensure the security of communication. It is only used to generate and distribute keys and does not transmit any substantive messages.

Application Level: Based on single-photon QKD technology, QKD uses decoy BB84 and other protocols and has been applied in finance, government, national defense, power and other fields. China is in the forefront of QKD applications, and many places (such as Hefei, Wuhu, and Jinan) implement metropolitan QKD networks using optical fibers. In March 2018, research and development achievements of field-wavelength division multiplexing at home and abroad, namely, quantum key distribution and 3.6 T bit/s classical communication of the commercial WDM backbone network, were published in international authoritative journals, such as "Optics Express". In July 2018, the "Physical Review Letters" were published in the form of "editorial recommendation", in which 18 optical quantum bits are entangled. The interstate quantum key distribution (IQKD) was completed by the Quantum Science Experiment Satellite in December 2018. The 1000-kilometer-scale two-way quantum entanglement distribution (QEDD) was realized by the Quantum Science Experiment Satellite in January 2019.

6.11   Hot Word 11: Internet of Vehicles

Basic Definition: The concept of the Internet of Vehicles is extended from the Internet of Things. The Internet of Vehicles is a large system network that can realize intelligent traffic management, intelligent dynamic information service and vehicle information exchange between vehicles and X (X: vehicles, roads, pedestrians, and the internet) in accordance with the agreed communication protocols and data interaction standards and based on the intra-vehicle network, inter-vehicle network and mobile Internet. The integrative network of vehicle intelligent control is a typical application of IoT technology in the field of transportation systems (quoted from "strategic alliance of technological innovation of vehicle network industry").

　　Application Level: The innovation of the Internet of Vehicles industry is becoming increasingly active and was in a period of strategic opportunity before the outbreak. Vehicle networking is a large information exchange network; its complexity requires cross-border cooperation between traditional automobile enterprises and internet and electronic communication enterprises. In 2018, the cooperation between major automobile companies and internet giants has become increasingly close. Cooperative development has become the mainstream, and the automobile network ecosystem has materialized. In terms of standards, the 3GPP R14 standard of LTE-V2X was officially issued in 2017; the 3GPP R15 standard of LTE-eV2X was formally completed in June 2018; and the 3GPP R16 + standard of 5G-V2X was launched in June 2018. Currently, Beijing, Shanghai, Chongqing, Wuhan Hubei and other cities have built Internet of Vehicles test demonstration zones and actively promote the testing and verification of various semi-closed roads and complex open roads.

6.12   Hot Words 12: 6G

Basic definition: 6G is the next generation mobile communication network after 5G. Compared with the 5G network, the 6G network has a higher network rate, lower communication delay and wider depth coverage. 6G will fully share ultra-high frequency radio spectrum resources, such as the millimeter wave, terahertz, and visible light, and integrate technologies such as ground mobile communication, Satellite Internet and microwave networks to form an integrated green network with all things involved in group cooperation, data intelligent perception, real-time security assessment, and space-ground integration coverage.

Application Level: Currently, 5G has only completed the first stage of standardization (R15), and 6G has not received a unified consensus in the international community. Many countries are looking forward to the future of 6G. Bell Laboratory of the United States believes that 6G should pay more attention to "human needs" and cover the whole earth and even space; communications experts in China believe that 6G should have three characteristics: full-spectrum, full-coverage and full-application [17, 18]. The coverage of existing mobile communications on the edge of a network, seas, deserts and other areas is relatively weak. Without sufficient sharing of spectrum resources, the realization of a full coverage 6G network of space-ground integration is facing greater challenges.

6.13   Hot Words 13: Satellite Internet

Basic Definition: Satellite Internet is the use of communication satellites in low, medium and high earth orbits to form a global Internet and is an important supplement to the existing ground mobile Internet. Satellite Internet can provide high-quality network and data transmission for specific areas, such as sea, remote mountainous area, desert, Gobi, high-altitude area and their industries. High-throughput Earth Satellite Networking can effectively build an integrated Space-ground integration network that serves the global scope.

Application Level: In April 2017, the first Chinese Ka-band multi-beam broadband high-throughput satellite (ZhongXing 16) was successfully launched with a total communication capacity that exceeds 20G bit/s. In December 2018, the Hongyun-1 satellite, which was used to establish a satellite constellation, was successfully launched. China plans to launch 156 Hongyun series satellites in 2022 and build a low-orbit communication satellite constellation to provide Internet services for various terrestrial terminals, improve Internet access in remote areas of China, and initially form the Satellite Internet. Other countries are also planning similar satellite constellation plans and have made corresponding preparations in terms of technology and funding. China is cooperating with other countries to explore and jointly develop satellite Internet for a wide area of the world.

6.14   Hot Words 14: Ocean Internet

Basic Definition: Ocean Internet refers to a network system that can provide access and data services similar to the Terrestrial Internet in a marine environment, which accounts for 71% of the global surface area. Underwater networks usually use submarine optical cables, acoustic waves and lasers as transmission media, while on the water surface, land mobile communications, satellite communications and wireless ad hoc networks are the main means of communication.

Application Level: More than 400 submarine optical cables exist in the world, and 99% of international cross-sea communication primarily depends on submarine cables. All countries in the world realize that submarine cables are the foundation of the national information development strategy and are investing and building submarine optical cable systems. However, China has mastered the technology of submarine optical cable self-manufacturing and urgently needs to develop a submarine cable fleet and a large number of submarine construction equipment. Because underwater acoustic waves, laser waves and radio waves have short communication distances and slow speeds, they are not suitable for large-scale applications. Global maritime mobile communication coverage is weak and satellite communication is expensive, which are shortcomings of marine network development. China has a vast ocean area. The development of the Ocean network has high economic value and strategic significance in the application of ocean transportation, oil exploration, marine environment monitoring and seabed exploration.

6.15 Hot Word 15: Open Source Systems

Basic Definition: An open source system refers to a system whose source code, design documents and other related documents are open to users instead of developers. Users can obtain the source code or design drawings according to different open source protocols and modify them. Open source systems respect the authors' property rights, and copyright is protected by law. Open source systems include open source software and open source hardware, which are part of the open source culture.

Application Level: Open source software system began to prevail in the 1990s. The technology system and open source protocols are relatively mature, such as GPL, MPL, BSD and other open source license agreements. Compared with open source software, open source hardware remains in its infancy and has not formed a mature technology development system, and fewer users participate in the development of open source hardware. The implementation of an open source system started relatively late in China. The open source ecosystem is not mature, the awareness of copyright is weak, and the open source culture requires further popularization. The development of an open source system in China can reduce the cost of system design, shorten the development cycle and improve the ability of self-control.

References

1. Center for Electronics and Information Studies, Chinese Academy of Engineering. Research on the development of electronic information engineering technology in China 2017 [M]. Beijing: Science Press, 2018.
2. Shaohua Yu. The process of digitalization and networking intellectualization accelerates the network communication technology with ten characteristics [N]. China Electronic News, 2018-08-14(7).
3. Shaohua Yu. The Seven Technical Walls of Network Communications and the Primary Exploration of Trends [J]. Study on Optical Communications, 2018, (5): 5-11, 28.
4. Shaohua Yu. A New Paradigm of Future Network: Net-AI Agent and City-AI Agent [J]. Study on Optical Communications, 2018, (6):5-14.
5. Shaohua Yu. Ten characteristics of network communication technology [N]. People's Posts and Telecommunications News, 2018-08-16(5).
6. Coordination Bureau of Operation Monitoring, Ministry of Industry and Information Technology. Operation of Electronic Information Manufacturing Industry in 2018 [EB/OL]. 2019. http://www.miit.gov.cn/n1146312/n1146904/n1648373/c6635637/content.html.
7. Shaohua Yu. The layout of "cyber power" and "facilitating faster and more affordable" has greatly improved broadband capacity [N]. China Electronic News, 2017-03-07 (5).
8. ITU. Measuring the information society report 2018 [R/OL]. 2018. https://www.itu.int/en/ITU-D/Statistics/Pages/publications/misr2018.aspx.
9. Coordinating Bureau of Operation Monitoring, Ministry of Industry and Information Technology. Statistical Bulletin of Communications Industry 2018 [EB/OL]. 2019. http://www.miit.gov.cn/n1146312/n1146904/n1648372/c6619958/content.html.
10. Xiaohu Yu. Communication network convergence and technique revolution[J]. Scientia Sinica (Informationis), 2017, (1): 148-152.
11. Jianfei Hu, Ning Ding. Application of Internet of Things in Intelligent Fire Fighting [J]. Electronic World, 2019, 560 (2): 194, 196.
12. Xiaoguang Shi, Song Luo. Pain Point, Hot Point and Power Point of Edge Computing of Industrial Internet [N]. People's Posts and Telecommunications News, 2018-12-28 (4).
13. China Academy of Information and Communications Technology. White Paper: Report on the Development and Application of Quantum Information Technology (2018) [R/OL]. 2018. http://www.caict.ac.cn/kxyj/qwfb/bps/201812/P020181221551983008723.pdf.
14. Wen Kai. Fully promote the deep integration of 5G and real economy [N]. People's Posts and Telecommunications News, 2018-06-25.
15. Ying Du, Hao Zhu, Hongmei Yang, etc. Review of 5G mobile communication technology standard [J]. Telecommunications Science, 2018, (8): 2-9.

© China Science Publishing & Media Ltd (Science Press) 2020

Center for Electronics and Information Studies, Chinese Academy of Engineering, *Network and Communication*, https://doi.org/10.1007/978-981-15-4596-2

16. World Bank. 2009 Information and communications for development: Extending reach and increasing impact [J]. World Bank Publications, 2010, 30 (1): 1-5.
17. Liang Chen, Shaohua Yu. Preliminary Study on the Trend of 6G Mobile Communication. Study on Optical Communications [J], 2019, (4): 1-8.
18. Liang Chen, Shaohua Yu. Preliminary Study on the Key Technologies of 6G Mobile Communication [J]. Study on Optical Communications, 2019, (5): 1-7.

Printed in the United States
By Bookmasters